STECK-VAUGHN

GED

Science

EXERCISE BOOK

ROSE MARIE BIDDIER

About the Author

Rose Marie Biddier has served as a science consultant to National Public Radio. She is known for her successful experiential teaching techniques and has trained teachers on three continents. She has adapted philosophy of science curriculum for non-English speaking immigrants and has written Earth and physical science curriculum for students with specific learning and reading deficiencies. She has developed physical science criterion-referenced test questions for Prince George's County Public Schools in Maryland where she teaches at Bladensburg Senior High School.

Staff Credits

Executive Editor: Ellen Northcutt
Supervising Editor: Tim Collins
Design Manager: John J. Harrison
Cover Design: Rhonda Childress

ISBN 0-8114-7370-8

Printed in the United States of America

1 2 3 4 5 6 7 8 9 DBH 99 98 97 96 95

Contents

To the Learner

The *Steck-Vaughn GED Science Exercise Book* provides you with practice in answering the types of questions found on the actual GED Science Test. It can be used with the *Steck-Vaughn GED Science* book or with the *Steck-Vaughn Complete GED Preparation* book. Cross references to units in the those books are supplied for your convenience on exercise pages 4–34. This exercise book contains both practice exercises and two complete simulated GED tests.

Practice Exercises

The GED Science Test examines your ability to understand, apply, analyze, and evaluate information in four science areas. The practice exercises are divided into the same four content areas by unit. Biology examines living things and how they interact with each other and their environment. Biology covers plant and animal biology, including human body systems. Earth science examines Earth and the space around Earth. Chemistry is the study of matter and changes in matter. The fourth area, physics, is the study of the interrelationships of matter and energy, including heat, light, sound, electricity, magnetism, atomic reactions, and motion.

Simulated Tests

This workbook contains two complete full-length Simulated GED Science Tests. Each Simulated Test has the same number of items as the GED Test. In addition, each test provides practice with similar item types that are found on the GED Test. The Simulated Tests can help you decide if you are ready to take the GED Science Test. To benefit most from the Simulated Tests, take each test under the same time restrictions as you will have for the actual GED Test. For each test, complete the 66 items within 95 minutes. Space the two examinations apart by at least a week.

Question Types

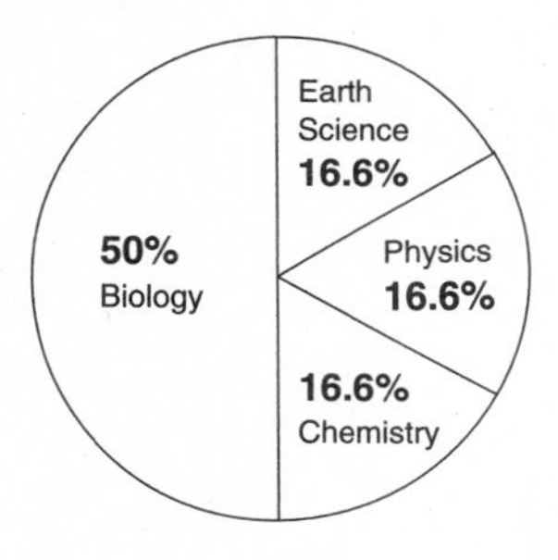

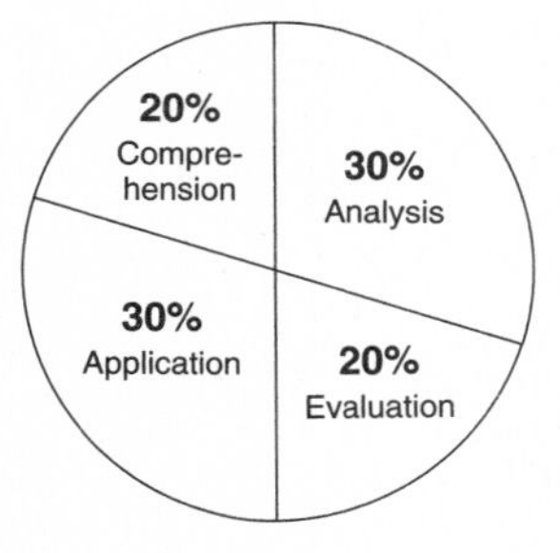

The GED Science Test is divided into four content areas: **biology, Earth science, physics,** and **chemistry.**

Short reading passages usually have only one question following them. Longer passages have several related questions. Half of the passages and related items (33 questions) are about biology. From your life experience, you may easily relate to these items. The other half is divided approximately equally with 11 questions each in Earth science, chemistry, and physics.

All of the questions on the GED Science Test are multiple-choice. You will not be tested on your knowledge of science, but rather on your ability to understand, apply, and analyze science concepts. Following is an explanation of the four types of questions that you will practice in this book and that are found on the GED Test.

1. **Comprehension** items require you to identify restated information or information that is paraphrased. They require you to summarize ideas or identify implications.
2. **Application** items require you to apply a rule and make a prediction of what would happen in a similar instance. They require you to use the information provided to solve a problem.

3. Analysis items require you to classify information. Sometimes you will be asked to distinguish or compare and contrast information presented.
4. Evaluation items test your ability to identify opinions and/or recognize assumptions. Other evaluation items ask you to identify cause and effect relationships.

Approximately 20% of the items are comprehension, 30% are application, 30% are analysis, and 20% are evaluation.

Illustrations

Approximately one-third of the items relate to a drawing, chart, map, or graph. Practice with graphics is essential to develop the skills to interpret information presented on the GED Science Test. Always read the title, key, and any other information associated with the illustration before answering any questions.

Answers

The answer section gives complete explanations of why an answer is correct, and why the other answer choices are incorrect. Sometimes by studying the reason an answer is incorrect, you can learn to avoid a similar problem in the future.

Analysis of Performance Charts

After each Simulated Test, an Analysis of Performance Chart will help you determine if you are ready to take the GED Science Test. The charts give a breakdown by content area (biology, Earth science, chemistry, and physics) and by question type (comprehension, application, analysis, and evaluation). By completing these charts, you can determine your own strengths and weaknesses as they relate to the science area.

Correlation Chart

The following correlation chart shows how the sections of this exercise book relate to sections of other Steck-Vaughn GED preparation books. You can refer to these two books for further instruction or review.

CONTENT AREAS	Biology	Earth Science	Chemistry	Physics
BOOK TITLES Steck-Vaughn GED Science	Unit 1	Unit 2	Unit 3	Unit 4
Steck-Vaughn GED Science Exercise book	Unit 1	Unit 2	Unit 3	Unit 4
Steck-Vaughn Complete GED Preparation book	Unit 4, Biology	Unit 4, Earth Science	Unit 4, Chemistry	Unit 4, Physics

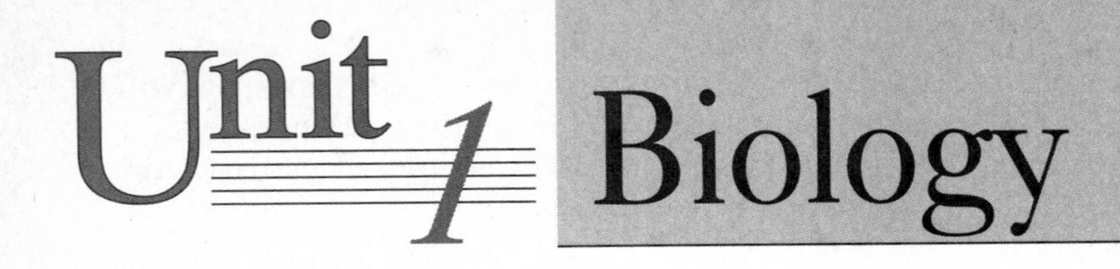

Unit 1 Biology

Directions: Choose the best answer to each item.

Items 1–5 refer to the following illustrations.

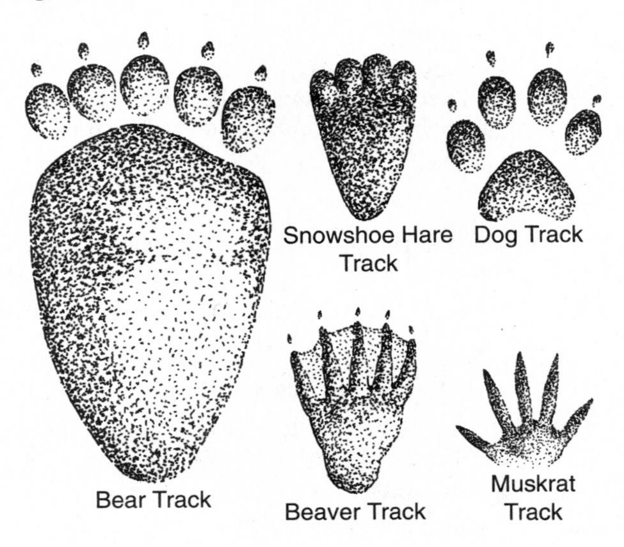

1. Which tracks show evidence of toenails?

   (1) bear, muskrat, beaver
   (2) bear, dog, beaver
   (3) snowshoe hare, dog, muskrat
   (4) beaver, muskrat
   (5) snowshoe hare, muskrat

2. The animal whose track most closely resembles a human footprint is the

   (1) beaver
   (2) bear
   (3) snowshoe hare
   (4) dog
   (5) muskrat

3. The one characteristic of the beaver's track that would most assist in building its house midstream is the

   (1) number of toes
   (2) length of toes
   (3) webbing between the toes
   (4) absence of pads
   (5) absence of toes

4. From observation of the muskrat tracks, the most likely place to find its burrow would be in

   (1) junctions of tree branches
   (2) midstream
   (3) a factory chimney
   (4) soft soil
   (5) high hollows of trees

5. Assuming the tracks are drawn to scale, which of the following statements is supported by the illustrations?

   (1) Snowshoe hares live in forests.
   (2) Muskrats eat fish.
   (3) Beavers are larger than muskrats.
   (4) Beavers are related to dogs.
   (5) Snowshoe hares have five toes on their front feet.

See Also

Science Text	Unit 1
Complete Preparation	Unit 4, Biology

Items 6–10 refer to the following passage.

Animals are killed chiefly for their meat value. Leather is a by-product derived from the skins of those animals. To obtain lasting, flexible, attractive leather from hides involves a series of chemical processes. The skins are soaked in brine (salt water) to kill bacteria that naturally rot and decay the hides. They are then dried and scraped to remove any remaining fat, flesh, or dirt and soaked in a solution of slaked lime and sodium sulfide to loosen the hairs. A rinse of pancreatic extract and ammonia removes the lime.

The hides are then tanned using an extract of hemlock or oak tree bark called tannin, which prevents further decay by bacteria. Various dyes are used to color the leather, which is then oiled with castor bean or cod liver oil. When dry, lacquers or waxes are painted on the leather to give it a shine and to prevent moisture and bacteria from entering the leather.

6. All of the following statements can be verified by the passage except which one?

   (1) The preparation of leather from hides is a multistep process.
   (2) Without chemical treatment, the hides would naturally decay.
   (3) Some of the chemicals used to treat hides come from plants.
   (4) Many ancient people used leather garments; therefore, the processing is simple and easy.
   (5) Salt is one of the chemicals used to process hides.

7. Which of the following (raw material to product) would be processed most like leather?

   (1) pelts to fur coats
   (2) wood to paper
   (3) plant juices to ink
   (4) plant fibers to nylon
   (5) petroleum to gasoline

8. If the price for the chemicals used in processing leather increased significantly, all of the following increases would occur except

   (1) the cost of leather shoes
   (2) the use of cloth, plastics, and other products to make shoes
   (3) the cost of leather upholstery for cars and furniture
   (4) the use of shoes as clothing
   (5) the use of cloth and synthetics for handbags, luggage, and belts

9. The most likely reason that protests against the use of animal skins for leather are not as intense as protests against skins for fur coats is that

   (1) the skins for leather are a by-product from animals that have already been killed for their meat
   (2) people have always used leather; whereas, the use of furs is a recent development
   (3) the animals suffer more when killed for fur
   (4) the animals used for leather are not as beautiful as those used for furs
   (5) coats can be made of cloth; whereas, the manufacture of footwear requires leather

10. Very few people are allergic to animal skin; however, many individuals are allergic to leather. Which of the following explanations would most likely be the cause of the allergies?

   (1) In leather, the skin is not alive.
   (2) In leather, the hairs have been removed.
   (3) The allergy is due to the chemicals used in processing leather.
   (4) The allergy is psychological in that an animal had to die to obtain the leather.
   (5) An increase in bacteria activates the tannin.

<u>Item 11</u> refers to the following graph.

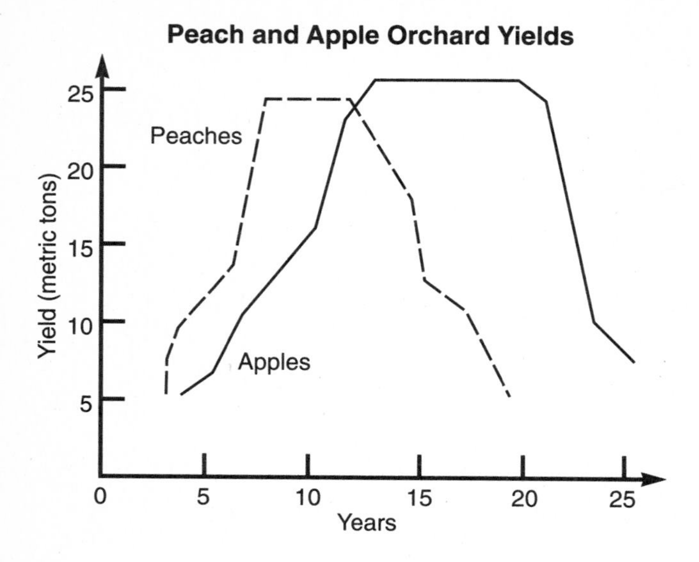

11. The graph gives information on all of the following considerations regarding fruit production <u>except</u>

(1) the approximate age of the trees before fruit production begins
(2) the approximate age when the trees reach peak production
(3) the approximate age of the trees when fruit production ceases
(4) information to indicate that peak production of peaches will fall off before that of apples
(5) data regarding the production of oranges in metric tons

<u>Items 12–18</u> refer to the following passage.

Pollination is the first step in a flowering plant's reproduction. In order to produce a seed that can become a new individual, the plant needs to flower. Delayed or premature flowering often results in diminished seed quality. In general, healthy seeds come from healthy plants.

Flowers contain male and/or female structures. The male structure produces pollen, which contains the sperm. The female structure produces ovules, which contain the eggs. Each flower of tomato and pepper plants contains both male and female structures. Melons, squash, and cucumbers have some flowers that are male and other flowers that are female on the same plant. Pussy willows and papayas have only male flowers on one plant and female flowers on another plant.

Regardless of where the male and female structures are located, there can be no fruit, and subsequently no seeds, unless the sperm in the pollen reaches the ovules. Wind, insects, bats, birds, moths, and butterflies assist plants in the transfer of pollen to the female structure. A flower's colored petals, scent, and nectar are devices for attracting these natural agents of pollination. In situations where the natural transfer agents are missing, humans can assist the plant.

12. Which of the following is <u>not</u> considered a natural agent of pollination?

(1) humans
(2) moths
(3) insects
(4) bats
(5) wind

13. The flowers of wild grasses are often small and have no petals, scent, or nectar. The agent of pollination for most wild grasses is probably

(1) humans
(2) moths
(3) insects
(4) bats
(5) wind

14. The label on a fruit-producing plant at a nursery says the plant won't produce fruit if there is only one in your yard. What can you infer about how this plant reproduces?

(1) Male and female flowers grow on separate plants.
(2) Both male flowers and female flowers grow on the same plant.
(3) All the flowers have both male and female structures.
(4) Humans must pollinate the plant.
(5) The plant doesn't reproduce by flowers.

15. The function of flowers is to

(1) increase the beauty of the plant
(2) notify people when the plant is mature
(3) decrease the excess nitrogen absorbed by the plant
(4) increase the nutrient levels needed for reproduction
(5) produce the reproductive elements necessary for seed formation

16. The best assurance that seeds will become healthy new plants is that

(1) the flowers are complete
(2) pollen has reached the female element
(3) the fruits for nourishment are large
(4) the female element produces many ovules
(5) the parent plant is healthy

17. The most likely place humans would need to assist in pollination would be

(1) on large farms of one crop type
(2) with nonflowering plants
(3) in desert or drought conditions
(4) in a commercial greenhouse
(5) in a backyard flower garden

18. Most moths fly only at night. They are guided by their sense of smell and by their sense of sight. Moth-pollinated flowers are most likely to be

(1) large and brightly colored
(2) large, white, and fragrant
(3) small, white, and odorless
(4) small, brightly colored, and fragrant
(5) open only during the day

Items 19–22 refer to the following illustration.

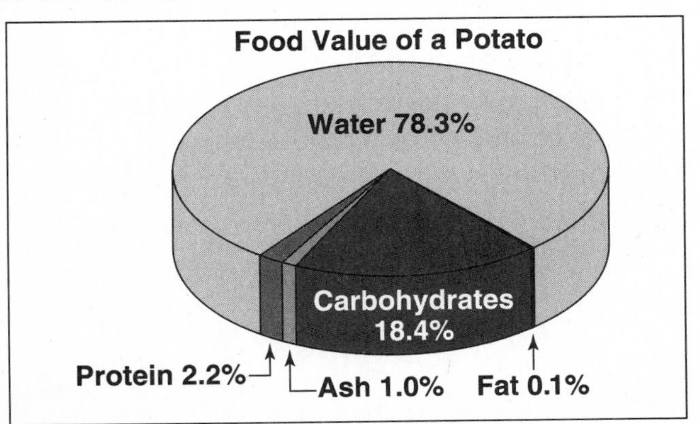

19. The reason potatoes are not considered high in calories is that they are

(1) mostly protein
(2) high in water content and low in fat
(3) low in ash and water
(4) vegetables and not fruits
(5) low in food value

20. The small dents on the surface of potatoes are called eyes. Each eye can sprout by producing a bud which can become a new plant. Potato farmers most likely start new plants by

(1) spreading seeds
(2) planting runners from mother plants
(3) placing the flowers from female plants in water to root
(4) cutting a potato into sections each having at least one eye, and then planting the sections
(5) plowing under the stem and leaves of old plants

21. Which of the following is not likely to happen to the food value of a potato when it is cut up and fried in hot oil to make French fries?

(1) fat content increases
(2) water content decreases
(3) protein content stays the same
(4) carbohydrate content increases
(5) ash content stays the same

22. If all the water is removed from a 100-gram potato to make dried potato flakes, how much will the flakes weigh?

(1) 2.2 grams
(2) 18.4 grams
(3) 21.7 grams
(4) 78.3 grams
(5) 81.6 grams

Items 23–25 refer to the following information.

Keys are used by biologists to classify and identify plants and animals. When using a key, a person always starts at the top and answers yes or no to the qualifying information.

If attempting to identify Fish E, a person would first ask if the fish is saucer-shaped. Upon answering no, the second question is asked. Are the stripes on the body of the fish vertical and narrow? The answer is yes, and the arrow identifies the fish as a zebra fish. If a person knows the name of a fish and wishes to determine if a sample is that fish, identification is positive when the person answers yes to the question with the arrow pointing to the name and no to all the above questions. Following is an example of the process.

(1) Is the fish saucer-shaped?
If yes → stingray
If no, go to 2.

(2) Are the body stripes vertical and narrow?
If yes → zebra fish
If no, go to 3.

(3) Is the mouth near the bottom of the head?
If yes → scorpion fish
If no, go to 4.

(4) Is there a lateral horizontal stripe parallel to the dorsal (top) fin?
If yes → weever fish
If no, go to 5.

(5) Are there vertical stripes on the caudal (tail) fin?
If yes → stonefish

Fish A:

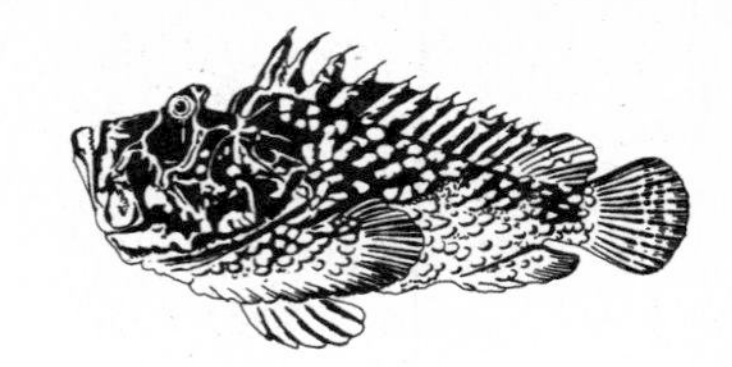

Fish B:

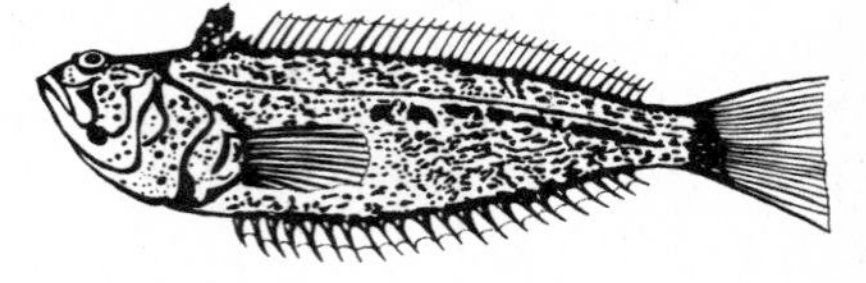

Fish C:

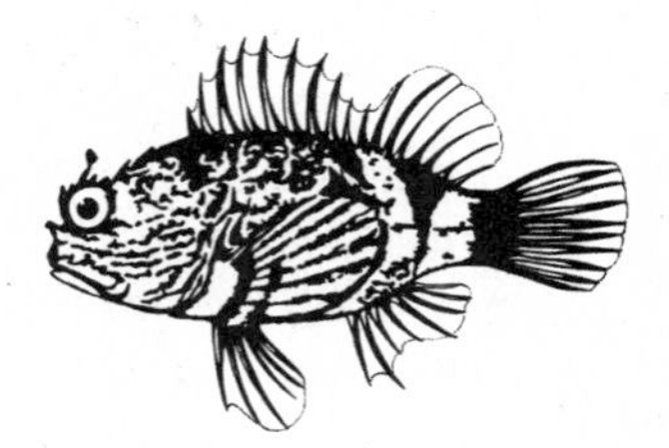

Fish D:

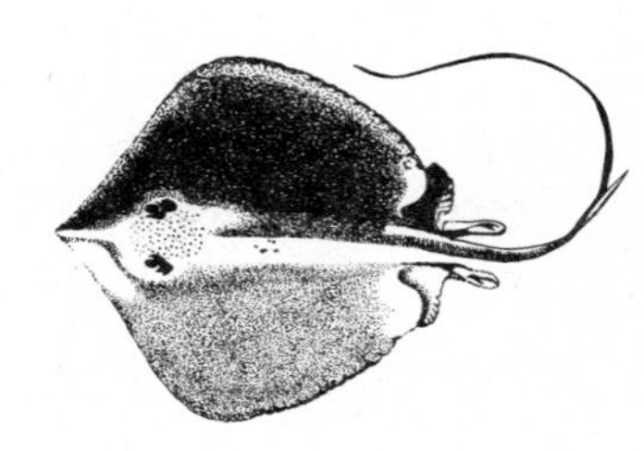

Fish E:

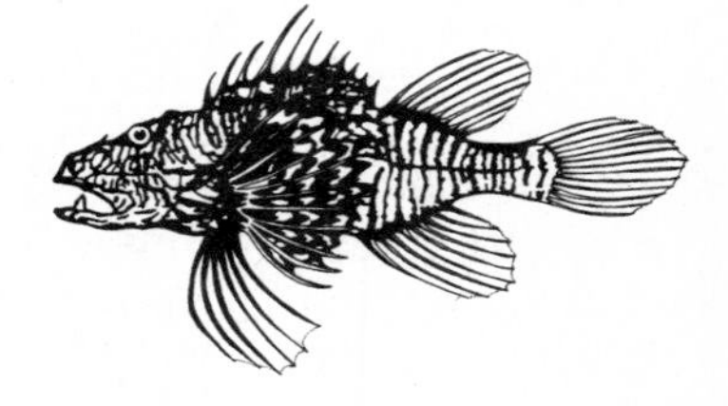

23. Using the classification key for the venomous fish pictured, Fish D is identified as a

(1) stingray
(2) zebra fish
(3) scorpion fish
(4) weever fish
(5) stonefish

24. Using the classification key for the venomous fish pictured, Fish B is identified as a

(1) stingray
(2) zebra fish
(3) scorpion fish
(4) weever fish
(5) stonefish

25. Using the classification key for the venomous fish pictured, which letter identifies a scorpion fish?

(1) A
(2) B
(3) C
(4) D
(5) E

26. A weed is a plant growing where it is not wanted. Which of the following plants is not a weed in the situation described?

(1) a pasture covered with buttercups that contain an irritating juice that cattle dislike
(2) dandelions and quackgrass in the manicured lawn of a private home
(3) bindweed strangling a farmer's cornstalks
(4) water hyacinths locking a waterway used by canoeists
(5) Queen Anne's lace beautifying the roadside of an interstate highway

27. Many United States weeds were brought here as seeds by colonists for use as medicines, seasonings, pest control, cosmetics, scents, and dyes. As the production of chemical products was taken over by large companies, many of these prized plants were neglected and came to be viewed as weeds.

Which of the following statements best represents the main idea of the paragraph above?

(1) Once a weed, always a weed.
(2) All of America's weeds are foreign born.
(3) Weeds are America's best source of chemicals.
(4) Many of today's weeds were the prized plants of colonists.
(5) Weeds have many uses in today's world.

28. Many methods are used to control weeds. All of the following methods of weed control are safe for a food crop except

(1) quarantine laws with inspection at ports of entry
(2) inspection of commercial seeds for limits on the kind and percent of weed seeds contained
(3) hoeing and cultivating
(4) placing plastic sheeting, heavy paper, or mulch around plants
(5) heavy spraying with unregulated chemicals that poison weeds

Items 29–32 refer to the information below.

Uses of the Camel by Desert People

Transport	carry heavy loads plow fields turn water wheels transport humans
Food	milk, butter, cheese, and meat
Fibers	camel hair clothing and blankets
Skin	leather shoes, water bags, and tents
Bones	carved utensils and jewelry
Droppings	fuel for warmth and cooking

29. In some areas of Africa and Asia, the survival of desert people is

(1) threatened by the overpopulation of camels
(2) threatened by competition with camels
(3) dependent on the products and services of camels
(4) dependent on the elimination of camels
(5) entirely independent of camels

30. Contrary to common belief, camels do not carry water in their humps. The humps contain fat, which in time of limited food supply is turned into water and sugar. A camel's hump is likely to be largest when the

(1) female is pregnant
(2) male is used to plow fields or carry loads
(3) water supply is abundant
(4) food supply is abundant
(5) camel is young

31. The fact that camels sweat very little increases their

(1) water intake needs
(2) need for salt
(3) ability to spit when angered
(4) ability to survive in dry climates
(5) reproductive capacity

32. Camels have three eyelids over each eye. A likely function of this adaptation is to

(1) keep the eyes dry
(2) prevent sand from entering during sandstorms
(3) sleep better at night
(4) see better in the daytime
(5) appear blind to its enemies

33. Desert plants must have structures to keep water in; whereas marshland plants must have structures to keep water out. Which of the following adaptations would serve both desert and marshland plants?

(1) spines and thorns
(2) thin, transparent skin
(3) thick, tough skin
(4) big flowers
(5) many leaves

Items 34–37 refer to the following illustration.

Punnett Square Showing the Cross Tt X Tt

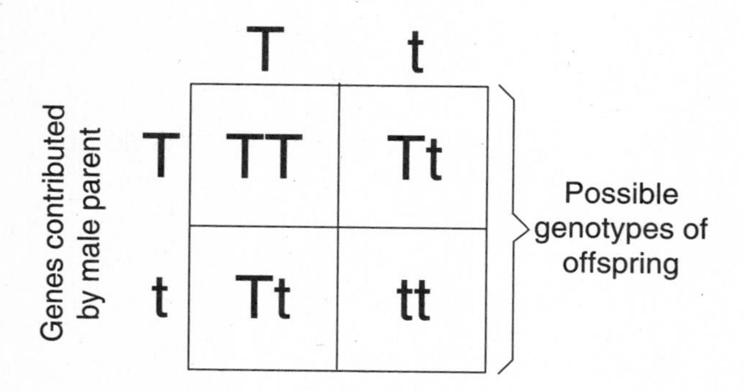

34. The parents in the genetic cross shown in the diagram have the genotypes

(1) Tt and Tt
(2) TT and tt
(3) Tt and TT
(4) tt and tt
(5) TT and TT

35. In the cross shown, the gene T is dominant over the gene t. If the cross produces 12 offspring, how many of them are likely to show the dominant trait, T?

(1) 10
(2) 9
(3) 8
(4) 3
(5) 1

36. What are the most likely results of a cross between parents with genotypes Tt and tt?

(1) all TT
(2) all tt
(3) all Tt
(4) half Tt and half tt
(5) one-fourth tt and three-fourths Tt

37. Many inherited human diseases, such as Tay-Sachs and sickle-cell anemia, are caused by a recessive gene. A person born with one copy of the gene is a carrier. A person born with two copies of the gene has the disease.

Which of the following couples has a chance of having a child with sickle-cell anemia?

(1) a man who is a sickle-cell carrier and a normal woman
(2) a normal man and a woman who is a sickle-cell carrier
(3) two sickle-cell carriers
(4) a normal woman and a man who has sickle-cell anemia
(5) two normal people

Items 38–40 refer to the following information.

The only sugar the body cells can burn for energy is glucose. The sugar in foods we eat is rarely glucose. Some sugars are listed below with their sources.

Single Sugars
- glucose—blood plasma
- fructose—fruits

Double Sugars
- maltose—grains and seeds
- sucrose—beets and cane (table sugar)

In the digestive process, the body breaks down double sugars and then changes all simple sugars to glucose before they are transported to the cells for energy.

Human Digestion of Sucrose

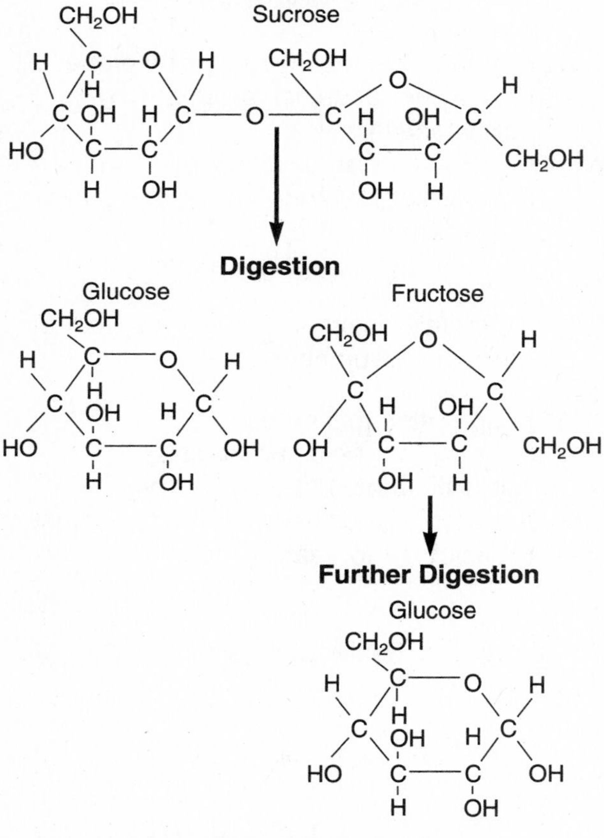

Sucrose

Digestion

Glucose

Fructose

Further Digestion

Glucose

38. All of the following statements are suggested in the information except

    (1) sucrose is a double sugar
    (2) sucrose must be broken down before body cells can use the energy
    (3) the sugar in fruits goes directly to cells without being changed to glucose
    (4) beets contain sugar
    (5) the sugar in grains is called maltose

39. Which of the chemicals listed below is most likely a sugar?

    (1) hydrochloric acid
    (2) ammonia
    (3) mannose
    (4) magnesium hydroxide
    (5) glue

40. When manufacturers of candy and heavy syrup products list a package's ingredients, they use the chemical names for sugar. Many people who wish to control body weight are advised to limit sugar intake. Most people do not recognize the chemical names of sugars. The most likely result of this gap of knowledge by the general public is that many consumers

    (1) assume the products have no sugar
    (2) buy the products because they sound scientific
    (3) do not purchase the product because they are unsure of the ingredients
    (4) become ill from the products
    (5) refuse to buy the products until the Food and Drug Administration forces the manufacturers to use common names

---

Items 41–42 refer to the following information.

Microorganisms living in the stomachs of ruminants (animals that chew a cud, such as cows) change the sugar of cellulose to glucose that the ruminant can then digest. Cellulose is found in grass and many plant fibers. Humans are unable to obtain glucose from cellulose.

41. From this information you can infer that

    (1) cows have a more highly developed digestive system than humans
    (2) humans must take vitamins in order to digest glucose
    (3) cellulose is the sugar used by the cells of ruminants to obtain energy
    (4) the microorganisms that digest cellulose do not live in the human digestive system
    (5) ruminants cannot digest fructose, sucrose, or maltose whereas humans can

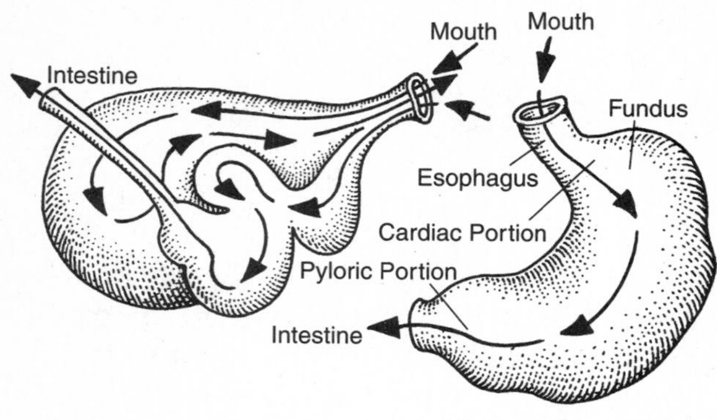

42. When contrasting the ruminant stomach to the human stomach, the ruminant stomach has all of the following except

(1) larger chambers
(2) more chambers
(3) a longer food route
(4) more than one food route
(5) a single pouch

Items 43–45 refer to the following passage.

Bacteria are helpful in the breakdown of dead organisms into simple molecules that can then be reused by new organisms. Some bacteria live in nodules on the roots of plants in the pea family and help replenish the soil with nitrogen essential to plant growth. People use bacteria which cause fermentation to preserve food such as cheese, vinegar, and sauerkraut for human consumption and silage for cattle. Bacteria are also used to purify water in sewage treatment plants and to produce insulin for diabetics in pharmaceutical laboratories.

Other bacteria are considered harmful. Some cause food spoilage and poisoning as in botulism and salmonellosis. Both plants and animals can become diseased by the invasion of certain bacteria. Leprosy, diphtheria, tuberculosis, gonorrhea, typhoid fever, and pneumonia are all bacterial infections. Black rot in cabbages and anthrax in sheep are also diseases caused by bacteria.

43. Which of the following statements best summarizes the information presented about bacteria?

(1) Bacteria are never harmful and are essential to human life.
(2) All bacteria are harmful.
(3) Both plants and animals can be harmed by bacteria.
(4) Some bacteria are harmful to humans, but others are helpful.
(5) Bacteria are more beneficial than harmful.

44. According to the passage, some bacteria are helpful to humans by

(1) killing the organism that causes cabbage rot
(2) producing insulin needed by diabetics
(3) purifying foods contaminated by salmonella
(4) curing anthrax in sheep
(5) poisoning the harmful members of the pea family

45. Which relationship is not similar to that of bacteria to pneumonia?

(1) cigarettes to lung cancer
(2) viruses to the common cold
(3) intestinal parasites to diarrhea
(4) salmonella bacteria to food poisoning
(5) accidents to death

Item 46 refers to the following illustration.

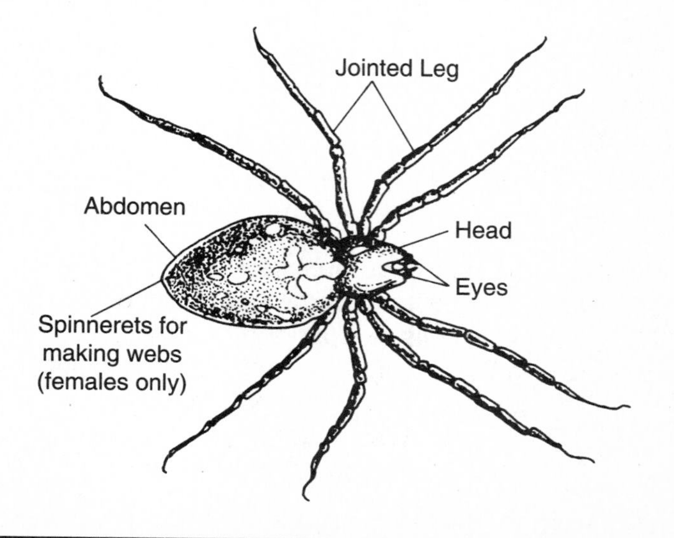

46. All insects have six legs and three body sections. The spider in the diagram is not an insect because it

(1) has eyes on top of its head
(2) has legs facing forward
(3) has pairs of legs each with 3 joint sections
(4) has an abdomen bigger than its head
(5) has eight legs and two body sections

Item 47 refers to the following paragraph.

Inherited genes determine what an individual may become; environment determines what an individual will become. A person may inherit from ancestors musical ability, but whether the person will be a musician is decided by factors in the environment.

47. According to the passage, the ability of an individual to excel at sports

(1) has no limit
(2) is only a result of inheritance
(3) is only a result of what a person eats
(4) is limited by inheritance but influenced by environmental factors
(5) will significantly increase with exercise

Items 48–49 refer to the following paragraph.

An advertisement for drinkable fiber says the drink includes the grain psyllium because it contains soluble fiber. Fiber assists in retaining sufficient water in the lower intestine to keep stools soft. This then helps a person to maintain regular and comfortable bowel movements.

48. Which of the following questions is least important in determining whether to use the product advertised?

(1) Does the individual need to include more fiber in the diet?
(2) Can the individual's diet be adjusted to include sufficient fiber from foods?
(3) Is psyllium grown by regular farming methods?
(4) If psyllium contains soluble fiber, does it hold as much water as insoluble fibers?
(5) Are any side effects caused by using this grain?

49. In the U.S. more money is spent on purgatives (medicine promoting evacuation of the bowels) than any other over-the-counter medicine. Many of the purchasers are elderly people who frequently experience irregularity. All of the following factors are likely contributors to this problem except the

(1) lack of sufficient exercise to stimulate the movement of food through the body
(2) lack of sufficient money to purchase high fiber fruits, vegetables, and grains
(3) breakdown of body processes so that the body does not function as well
(4) inability to know that a problem exists
(5) lack of interest in eating full regular meals that include a variety of food

---

Items 50–51 refer to the following paragraph.

At a health convention, a natural food salesperson warns, "Chemicals are the cause of cancer. If people would just buy natural foods instead of chemicals, our society would rid itself of this disease."

50. Not knowing that chemicals are the basic ingredients of all matter identifies the speaker as

(1) overeducated
(2) interesting
(3) misinformed
(4) unintelligent
(5) scientific

51. The basic fallacy (false assumption) in the sales pitch is that

(1) chemicals do not cause cancer
(2) chemicals are always safe
(3) natural substances are not chemicals
(4) chemicals changed by humans are safe
(5) chemicals are scientific

52. Oxygen is an odorless, colorless gas essential to life. In which of the following places is oxygen least likely to be found supporting life?

(1) ocean water
(2) topsoil
(3) hospitals
(4) the atmosphere
(5) metal alloys

Items 53–58 refer to the following information.

A large variety of creatures live on the continental shelves, which are covered with ocean water. The most numerous are the floating plankton. Zooplankton are microscopic animals, and phytoplankton are microscopic plants. Phytoplankton contain green chloroplasts and live near the surface of the ocean in order to receive enough light to synthesize their own food from carbon dioxide (CO_2) and water. Zooplankton have no green chloroplasts, cannot make their own food, and must eat phytoplankton. Fish and other swimming sea animals, called nekton, constantly supply carbon dioxide which the phytoplankton must have to synthesize food.

The nekton eat the plankton and other nekton. The benthos crawl on the sea floor, eating waste materials and dead plants and animals that have sunk to the bottom.

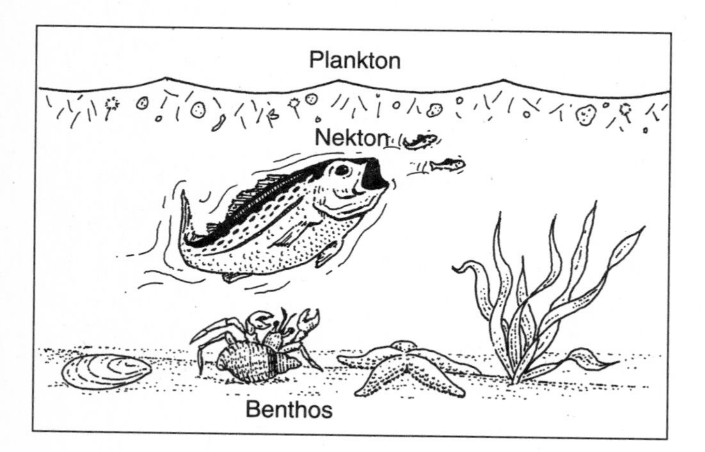

53. The ocean floor (benthos region) would be the natural home for all of the following sea creatures except
    (1) clams and scallops
    (2) whales and swordfish
    (3) sea snails and octopuses
    (4) shrimp and lobsters
    (5) starfish and sea urchins

54. If the phytoplankton in the ocean were to die off, which ocean animals would be affected?
    (1) none
    (2) zooplankton only
    (3) nekton only
    (4) benthos only
    (5) all ocean animals

55. In brown or red seaweed, the green chlorophyll is hidden by other chemicals. Giant kelp, a brown seaweed, often grows to 200 feet in length. Kelp has air bladders to help it float. The most likely reason this seaweed floats is to
    (1) trap phytoplankton
    (2) trap lobsters and fish
    (3) help them swim
    (4) keep them near the sunlight
    (5) stay near the air for easy breathing

56. The nekton are active swimmers that occupy specific levels within the ocean. The color of a fish depends on the level to which that species has adapted. Saltwater surface fish are usually blue or gray, mid-level fish are whitish or silvery, and very deep fish are very dark colors. Shoppers observe fish at the fish counter. Its skin is dark brown. This fish most likely
    (1) eats plankton
    (2) is a freshwater river fish
    (3) is a deep water fish
    (4) is not a true fish but a type of squid
    (5) is best served with butter

57. The most likely people to be interested in the specific levels occupied by different fish species would be
    (1) beach surfers
    (2) fishers
    (3) navigators of military ships
    (4) captains of ocean cruisers
    (5) Caribbean tourists

58. Many of the sun's rays are reflected off the ocean surface. The rays that enter the ocean water are absorbed by the water as thermal energy. Little light penetrates deep ocean water. Which statement below is most unlikely?
    (1) The surface water is hotter than water at the bottom.
    (2) The green plankton at the surface use some of the light.
    (3) It is not very dark in deep ocean water.
    (4) Deep water fish breathe with gills.
    (5) The deep water fish are often blind or have no eyes.

Answers are on page 76.

Unit 2 Earth Science

Items 1–5 refer to the following illustration.

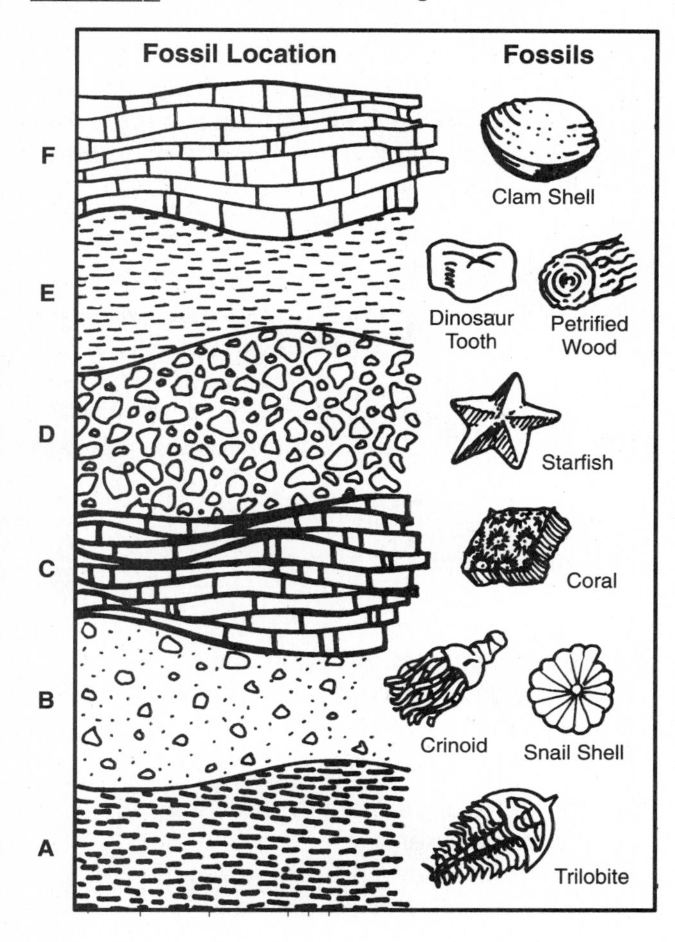

1. The law of superposition states that, for sedimentary rock, the layer below is always older. From the illustration, one can assume that the oldest fossil pictured is the
   (1) clam
   (2) coral
   (3) crinoid
   (4) snail
   (5) trilobite

2. Noting the fossils in the illustration, it is apparent that fossils are usually
   (1) plants
   (2) animals
   (3) sea organisms
   (4) land organisms
   (5) from hard parts of plants and animals

3. From the fossils imbedded in rock layer C, it is likely that the layer developed
   (1) from a river bed
   (2) deep on the ocean floor
   (3) in a desert
   (4) on forested land
   (5) on a shear mountain peak

4. Which of the following statements is contradicted by the fossil evidence in the illustration?
   (1) Bivalves such as clams evolved from snails.
   (2) Trilobites lived many millions of years ago.
   (3) Starfish evolved from crinoids.
   (4) Crinoids and snails first appeared in the oceans about the same time.
   (5) Clams were the first kind of animal to evolve in the oceans.

5. If a scientist found the fossils shown in the illustration by digging down at one spot, what could he or she infer about the geological history of that spot?
   (1) It was covered by the sea for a long time, then became dry land, then was covered by the sea again.
   (2) It was always covered by the sea, until recent times.
   (3) It was always dry land.
   (4) It was part of a continent that drifted north from the South Pole.
   (5) It has been covered by the sea for only a short time.

See Also
Science Text — Unit 2
Complete Preparation — Unit 4, Earth Science

Items 6–7 refer to the following illustration.

A maritime air mass refers to air that has spent time over a sea or ocean, while a continental air mass refers to air that has spent time over land.

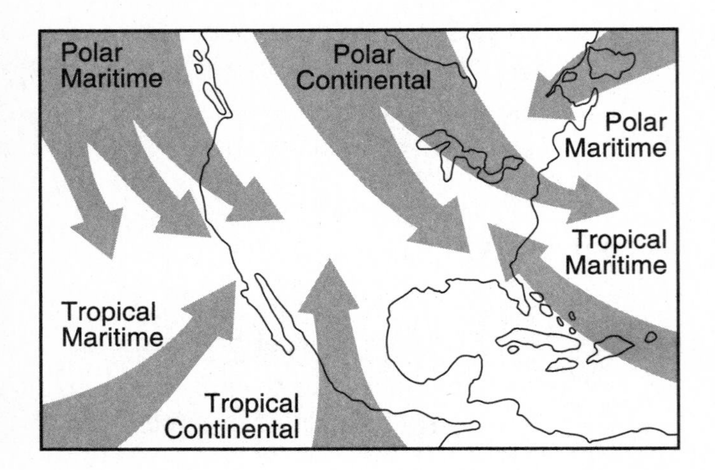

6. An air mass moving south from central Canada toward the midwest section of the United States would most likely be
   (1) wet and cool
   (2) wet and fast
   (3) humid and frigid
   (4) dry and cool
   (5) dry and hot

7. The hotter the air, the faster ocean water evaporates. Hurricanes develop near the equator when spinning air, holding large quantities of water, accelerates to speeds over 70 miles per hour. In which of the following air masses do most hurricanes form?
   (1) polar continental
   (2) polar maritime
   (3) tropical maritime
   (4) tropical continental
   (5) equal number in all types

Items 8–10 refer to the following information.

Soil is made up of decaying organic material and particles of rock, along with air and water. The rock particles range in size from almost-microscopic clay particles to gravel-size chunks. These components of soil tend to form layers. A certain kind of material dominates each layer.

The upper layers of a soil contain most of a soil's nutrients. These nutrients are required for plant growth. They come both from the soil's decaying organic matter and from the minerals in the soil's rock particles.

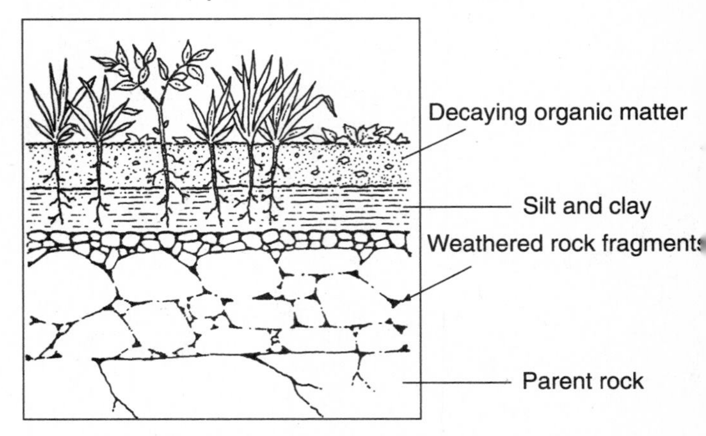

8. Earthworms eat decaying plant material in soil. You would expect to find the greatest number of earthworms
   (1) on the surface
   (2) in the uppermost layer of soil
   (3) in the layer of silt and clay
   (4) in the layer of weathered rock
   (5) in the layer of parent rock

9. Of the following places, where would you probably find the deepest soil?
   (1) in a desert
   (2) on a mountainside
   (3) near the Arctic circle
   (4) in a forest
   (5) where the parent rock is strong and hard

10. Which of the following soils is probably best for growing crops?
    (1) soil with a thick organic layer and no silt and clay
    (2) soil with thick layers of organic material and silt and clay
    (3) soil with no organic layer
    (4) soil with thin layers of organic material and silt and clay
    (5) soil with a thin organic layer and a hard-packed silt and clay layer

Items 11–15 refer to the following paragraph.

Usually sand accumulates in two ways—along beaches brought there by waves and in deserts brought or left there by wind. Sandstone is a rock formed from sand. Some kinds of limestone are formed from the shells and skeletons of dead sea organisms that sink to the bottom.

11. If a geologist digs down into rock in Indiana and finds a layer of limestone, the geologist can reason that

    (1) Indiana was once covered by an ocean
    (2) Indiana was once a desert
    (3) a giant tidal wave must have sent water to the center of the United States
    (4) an underground ocean once existed in that area
    (5) fish and sea creatures once lived on land

12. Sandstone is a sedimentary rock. Sedimentary rocks form in horizontal layers, like a sandwich. If you found sandstone in which each layer was a different color, this would be evidence that

    (1) the sand that formed the rock came from the ocean
    (2) a geological process sorted the sand by color before it became rock
    (3) animals burrowed in the sand before it became rock
    (4) the sand that formed the rock was blown there from different places
    (5) the sandstone was quarried in a certain way

13. Limestone (a rock) and clay (a soil) are mixed, then baked, and later ground to form cement. Cement is usually mixed with sand and gravel to make concrete. Limestone, soda ash, and sand are mixed, then heated, and molded or formed into glass objects. Modern societies used tremendous amounts of cement and glass for structures and objects of value. Which of the following structures or objects does not require limestone rock?

    (1) masonry blocks
    (2) glass bottles
    (3) windowpanes
    (4) sidewalks
    (5) plastic

14. Sandstone blocks can be cut and used instead of bricks for constructing buildings. Houses made of sandstone are often called brownstone houses. Sandstone is unusually absorbent but does not rust like metals nor rot like wood. However, smoke and dirt cling to sandstone and give it a dingy appearance. The most likely way brownstone houses are given a fresh clean look is by

    (1) painting with spray guns
    (2) washing with cleaning fluid
    (3) sandblasting with air guns
    (4) varnishing with hand brushes
    (5) covering with wallpaper

15. Rocks are made of minerals. Minerals that are hard are called abrasives and are used to smooth objects by rubbing. Sometimes the minerals are ground up prior to use. All of the following materials that are used to smooth objects contain minerals from rocks except

    (1) sandpaper
    (2) grinding wheels
    (3) hand lotion
    (4) scouring powders
    (5) fingernail emery boards

16. Some scientists believe that pollution of the atmosphere is causing the average temperature of the Earth's surface to increase and that it will continue to increase over a period of many years. They call this trend global warming. Which of the following provides the most convincing evidence for global warming?

    (1) an unusually warm summer in Indiana
    (2) two winters in a row in Norway with unusually warm temperatures
    (3) a three-inch rise in the level of the oceans caused by partial melting of the polar ice caps
    (4) a heat wave that breaks high-temperature records across much of the United States
    (5) five cities in the world recording average temperatures that are 1 to 3 degrees higher than last year's

Items 17–18 refer to the following information.

Beautiful minerals from rocks are worn as jewelry. Minerals called gems are usually quite scarce, which makes them valuable. Beautiful minerals that are plentiful are called semiprecious stones. The color of semiprecious stones or gems depends on the minerals and impurities they contain. Expensive jewelry often uses bright, colorful stones.

17. Rubies are red, sapphires are blue, and emeralds are green because

(1) they are scarce
(2) they are valuable
(3) the rocks in which they are found are very hard
(4) the minerals and impurities they contain are different
(5) they are gems and not semiprecious stone

18. Agates, which have bands of different colors, are a type of semiprecious stone. All of the following characterize agates except that they are

(1) small
(2) beautiful
(3) plentiful
(4) less valuable than gems
(5) colorful

19. Ores are rocks that contain one or more metals. Copper and iron are both metals. From this information one can deduce that

(1) copper and iron are strong
(2) iron is heavier than copper
(3) copper and iron come from rock materials
(4) copper and iron are valuable
(5) copper and iron are obtained from the same ore

20. Table salt is a mineral found in certain rocks and soils. Salt is also abundant in the oceans. Most table salt is obtained from inland underground mines. The most likely reason why salt is found in the oceans is that salt

(1) occurs only in rocks found at the bottom of oceans
(2) evaporated from rocks into the air
(3) dissolves in rainwater and is carried by rivers to the oceans
(4) is formed into rocks by animals in the oceans
(5) is a waste product dumped into the oceans by humans

Items 21–23 refer to the following information.

Each air molecule has very little weight, but air in the atmosphere is piled on top of Earth for many miles. At sea level, the weight of the air is 14.7 pounds on each square inch of surface. Rising above sea level decreases the column of air above any given point; thus the pressure is decreased at surfaces above sea level.

21. An instrument that measures air pressure very precisely could be used to determine

(1) temperature
(2) latitude
(3) elevation
(4) humidity
(5) wind speed

22. You can conclude from the drawing at the top of page 19 that

(1) the pressure of the atmospheric air decreases as elevation decreases
(2) the pressure of the atmospheric air increases as the elevation decreases
(3) the amount of air inside the balloon increases as the elevation decreases
(4) the amount of air inside the balloon decreases as the elevation decreases
(5) the changing volume of the balloon is independent of the atmospheric pressure

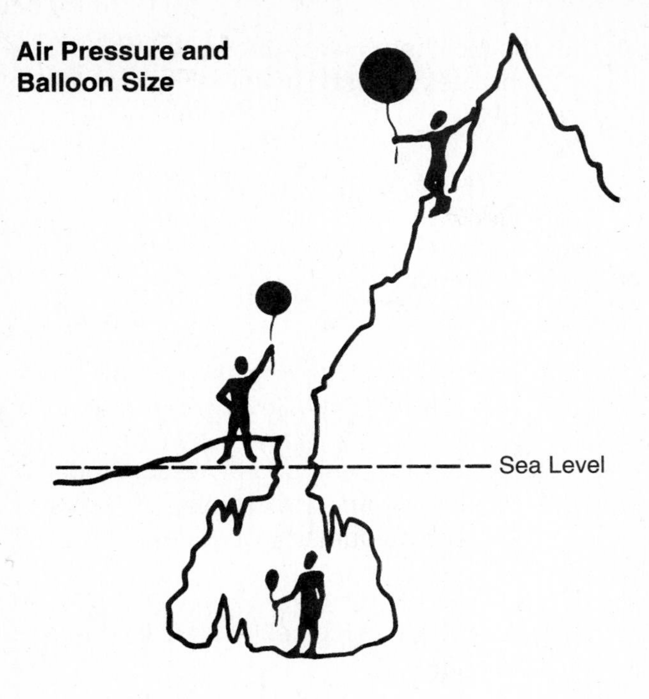

23. All of the following examples illustrate the same principle as the drawing <u>except</u> which one?

    (1) Yeast breads at higher elevations rise faster and higher than those at low elevations.
    (2) Deep sea divers must wear pressurized suits to maintain breathing and to prevent eardrums from bursting inward.
    (3) Astronauts walking in space or on the moon wear pressurized suits to prevent their bodies from bursting outward.
    (4) While touring the Empire State Building, a tourist's ears become painful on the elevator ride to the top and seem to pop on the ride down.
    (5) Trees are unable to grow on high mountain tops where air pressure is decreased.

Item 24 refers to the following illustration.

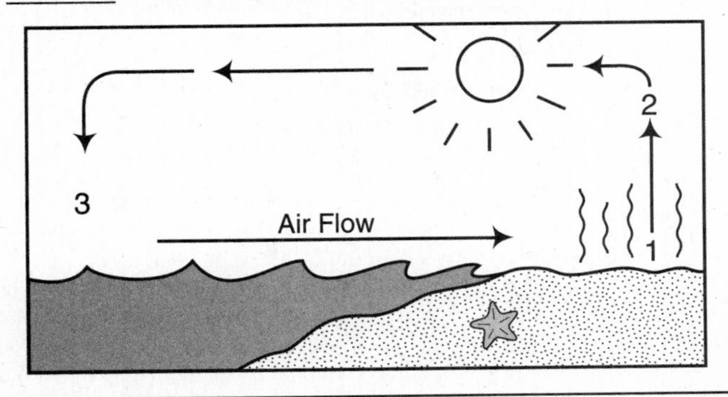

24. When air is hot, it rises. When air is cooled, it sinks, which causes wind. At sea coasts, all of the following conditions exist <u>except</u> which one?

    (1) The sand is hotter than the water.
    (2) During daytime, a cool breeze moves inland off the water.
    (3) The water temperature is lower than the temperature of the sand.
    (4) The sun sends more rays to land than it does to water.
    (5) The water cools the air above it, causing the air to sink.

Items 25–27 refer to the following information.

Average Yearly Earthquakes for Selected States

States	Earthquakes	High Intensity Quakes
Alaska	266	8
Arizona	24	1
California	830	30
Colorado	59	0
Hawaii	121	3
Montana	150	7
North Dakota	0	0
New York	52	2
Utah	101	3
Washington	141	10
West Virginia	6	0

25. Which of the statements below can be supported by data in the table?

    A. Earthquakes do not happen on the East Coast.
    B. Islands have almost no earthquake risk.
    C. California has the greatest number of earthquakes.
    D. None of Colorado's earthquakes were of high intensity.

    (1) A only
    (2) A and B
    (3) A and C
    (4) C and D
    (5) B, C, and D

26. Which of the following states has the greatest proportion of high intensity quakes to the total number of earthquakes?

(1) California
(2) Colorado
(3) North Dakota
(4) Utah
(5) Washington

27. The greatest number of earthquakes occur in those states

(1) in, near, or bordering the Pacific Ocean
(2) in, near, or bordering the Atlantic Ocean
(3) bordering Mexico or the Gulf of Mexico
(4) surrounding the Great Lakes
(5) in the central U.S. Great Plains

28. Erosion occurs when rocks or soil are moved from one place and deposited in another place. An agent of erosion is the force that has the energy to cause the rock or soil to move. Which of the following agents of erosion is the greatest factor in moving tropical desert sands and soils into piles called dunes?

(1) running water
(2) wind
(3) gravity
(4) ocean waves
(5) glaciers

Item 29 refers to the following drawing.

EARTH'S REVOLUTION AROUND SUN

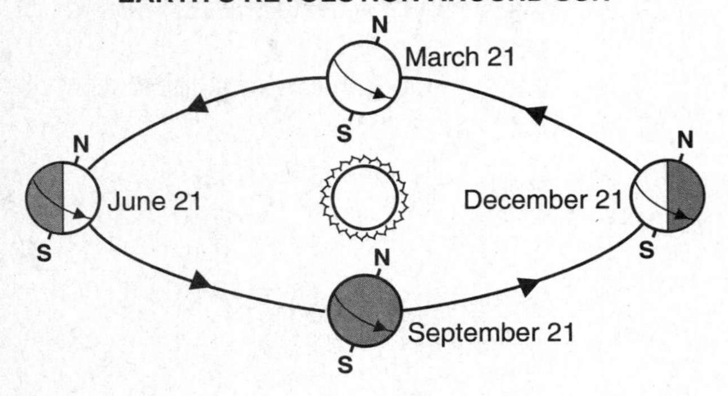

29. In the Northern Hemisphere, the date having the longest day and shortest night would be

(1) January 21
(2) March 21
(3) June 21
(4) November 21
(5) December 21

Items 30–31 refer to the following information.

Operating the devices used by people today requires substantial amounts of energy. Some of the ways people obtain this energy are listed below.

solar collection—using cells that collect heat from sunlight
burning fuels—obtaining heat from the remains of living material (wood, coal, petroleum, or natural gas)
nuclear fission—breaking apart certain atoms to release energy
geothermal energy—obtaining heat by digging deep into Earth's crust
tidal energy—using the natural movements of ocean water to operate turbines to obtain electricity

30. Which of the methods described is currently used to operate most land and air vehicles?

(1) solar cells
(2) burning fuels
(3) nuclear fission
(4) geothermal energy
(5) tidal energy

31. Which of the following sources of energy is most in danger of becoming exhausted?

(1) solar energy
(2) fossil fuels
(3) nuclear energy
(4) geothermal energy
(5) tidal energy

Items 32–35 refer to the following information.

When Earth makes one complete rotation on its axis, one day has passed. A year is measured by one revolution of Earth around the sun. During one year, 365 days, 5 hours, 48 minutes, and 46 seconds pass. To account for the extra hours and minutes, an extra day, February 29, is added every fourth year and on years ending in 00 and divisible by 400.

To make a calendar, one must account for the need of 365 days on regular years and 366 days on leap years. In 1931, the League of Nations held an international conference on calendar reform. One of the more than 500 plans submitted that was seriously considered for worldwide usage was the thirteen-month calendar below. Agreement could not be reached on any one calendar; thus still today, many different calendars are used throughout the world.

The Thirteen-Month Calendar

JANUARY

S	M	T	W	T	F	S
1	2	3	4	5	6	7
8	9	10	11	12	13	14
15	16	17	18	19	20	21
22	23	24	25	26	27	28

FEBRUARY

S	M	T	W	T	F	S
1	2	3	4	5	6	7
8	9	10	11	12	13	14
15	16	17	18	19	20	21
22	23	24	25	26	27	28

MARCH

S	M	T	W	T	F	S
1	2	3	4	5	6	7
8	9	10	11	12	13	14
15	16	17	18	19	20	21
22	23	24	25	26	27	28

APRIL

S	M	T	W	T	F	S
1	2	3	4	5	6	7
8	9	10	11	12	13	14
15	16	17	18	19	20	21
22	23	24	25	26	27	28

MAY

S	M	T	W	T	F	S
1	2	3	4	5	6	7
8	9	10	11	12	13	14
15	16	17	18	19	20	21
22	23	24	25	26	27	28

JUNE

S	M	T	W	T	F	S
1	2	3	4	5	6	7
8	9	10	11	12	13	14
15	16	17	18	19	20	21
22	23	24	25	26	27	28

LEAP DAY
June 29

SOL

S	M	T	W	T	F	S
1	2	3	4	5	6	7
8	9	10	11	12	13	14
15	16	17	18	19	20	21
22	23	24	25	26	27	28

JULY

S	M	T	W	T	F	S
1	2	3	4	5	6	7
8	9	10	11	12	13	14
15	16	17	18	19	20	21
22	23	24	25	26	27	28

AUGUST

S	M	T	W	T	F	S
1	2	3	4	5	6	7
8	9	10	11	12	13	14
15	16	17	18	19	20	21
22	23	24	25	26	27	28

SEPTEMBER

S	M	T	W	T	F	S
1	2	3	4	5	6	7
8	9	10	11	12	13	14
15	16	17	18	19	20	21
22	23	24	25	26	27	28

OCTOBER

S	M	T	W	T	F	S
1	2	3	4	5	6	7
8	9	10	11	12	13	14
15	16	17	18	19	20	21
22	23	24	25	26	27	28

NOVEMBER

S	M	T	W	T	F	S
1	2	3	4	5	6	7
8	9	10	11	12	13	14
15	16	17	18	19	20	21
22	23	24	25	26	27	28

DECEMBER

S	M	T	W	T	F	S
1	2	3	4	5	6	7
8	9	10	11	12	13	14
15	16	17	18	19	20	21
22	23	24	25	26	27	28

YEAR DAY
December 29

32. Which of the following features is not characteristic of the thirteen-month calendar?

(1) The year is divided into thirteen months.
(2) All months have the same number of days.
(3) Each year a person's birthday would fall on the same day of the week.
(4) Every month would always begin on a Sunday.
(5) There would be no leap year day.

33. If everyone agreed to use the thirteen-month calendar, this would most likely benefit persons in all of the following situations except persons

(1) working on monthly salaries
(2) calculating daily profits quarterly
(3) remembering the day of the week for a federal holiday or birth date
(4) working in foreign countries where the U.S. calendar is not used
(5) in the military responsible for worldwide operations

34. Which of the following is determined by the movements of Earth in relation to the sun, and not by social convention?

(1) length of a week
(2) when the year begins
(3) number of weekends in a year
(4) length of a year
(5) the day on which Thanksgiving falls

35. In the thirteen-month calendar, why is a "year day" needed every year?

(1) to celebrate a special holiday on the last day of the year
(2) because a year is a few hours more than 365 days long
(3) to celebrate the birthday of the inventor of the calendar
(4) because 365 divided by 13 has a remainder of 1
(5) because 13 is an odd number

Answers are on page 79.

Items 1–4 refer to the following chart.

Approximate Densities (Grams Per Cubic Centimeter) of Some Seasoned (Cured) Wood	
Pine	0.35–0.6 cm^3
Cedar	0.3–0.4 cm^3
Spruce	0.5–0.7 cm^3
Hickory, maple, oak	0.6–0.9 cm^3
Walnut	0.7 cm^3
Ebony	1.2 cm^3

1. The density of all objects is compared to the density of water, which is 1.0 gm/cm^3. Objects with densities less than 1 gm/cm^3 will float, and those with densities greater than 1 gm/cm^3 will sink in water. Which of the following woods will sink in water?

   (1) only ebony
   (2) ebony and walnut
   (3) hickory, maple, and oak
   (4) only pine
   (5) all woods

2. Hardwoods have high densities and softwoods have low densities. Cutting and nailing hardwoods is more difficult, but softwoods dent and scratch more easily. Which builder would least likely need the information provided by the table?

   (1) carpenter
   (2) apartment construction supervisor
   (3) road contractor
   (4) fishing boatwright
   (5) cabinetmaker

3. The information in the table is most important when purchasing wood by the

   (1) pound
   (2) tree
   (3) board foot
   (4) truckload
   (5) wood type

4. Pine and cedar are considered softwoods. Hickory, maple, oak, and walnut are considered hardwoods. Based on this information and the densities listed in the table, hardwood is best defined as a wood with a density

   (1) greater than 0.4 gm/cm^3
   (2) greater than 0.5 gm/cm^3
   (3) of 0.6 gm/cm^3 or greater
   (4) of 0.7 gm/cm^3 or greater
   (5) between 0.6 gm/cm^3 and 0.9 gm/cm^3

5. Despite the fact that many gases are invisible, odorless, and colorless, gases do occupy space, have weight, and are thus matter. Which of the following statements does not support the fact that Earth's atmosphere contains matter?

   (1) Heat shield tiles get hot during the reentry of the space shuttle craft.
   (2) A person can feel the wind.
   (3) Land animals can breathe.
   (4) Humans have walked on the moon and found no atmosphere there.
   (5) Airplanes can fly in the troposphere but not in outer space.

6. Heating increases the rate at which water evaporates because heating

   (1) makes the water molecules more likely to split apart into atoms
   (2) decreases the density of the water
   (3) makes the water molecules more likely to form bonds with each other
   (4) strips electrons from the water molecules
   (5) makes water molecules at the surface more likely to break away from the rest

See Also

Science Text	Unit 3
Complete Preparation	Unit 4, Chemistry

Items 7–9 refer to the following information.

As the number of known elements increased, it became cumbersome to write out their names when expressing chemical reactions. In the early 1800's, a Swedish chemist, Joens Jakob Berzelius, designed the current standardized system of scientific notation.

The capitalized first letter of the element's Latin name represents the element. If more than one element began with the same letter, a second small letter was attached.

C = carbon
Ca = calcium
Cl = chlorine
Cu = copper, Latin name cuprum

The letter symbol stands for one atom. To express two or more atoms, a numerical subscript is placed at the bottom right of that element's letter(s). The notation CO_2 means one atom of carbon with two atoms of oxygen.

7. The following word prefixes refer to numbers. Mon = 1, di = 2, tri = 3, tetra = 4, and pent = 5.

   Carbontetrachloride (a dry-cleaning fluid) would be written in scientific notation as

   (1) CCl_1
   (2) CCl_2
   (3) CCl_4
   (4) $CoCl_4$
   (5) $CuCl_4$

8. The notation CO_2 would be identified as

   (1) carbon dioxide
   (2) carbon monoxide
   (3) carbon pentoxide
   (4) carbon tetroxide
   (5) carbon trioxide

9. If H = hydrogen, O = oxygen, C = carbon, and hydrocarbon fuels contain only hydrogen and carbon, which of the following formulas is a hydrocarbon?

   (1) CH_4
   (2) $C_6H_{12}O_6$
   (3) H_2CO_3
   (4) $C_6H_{13}OH$
   (5) H_2COOH

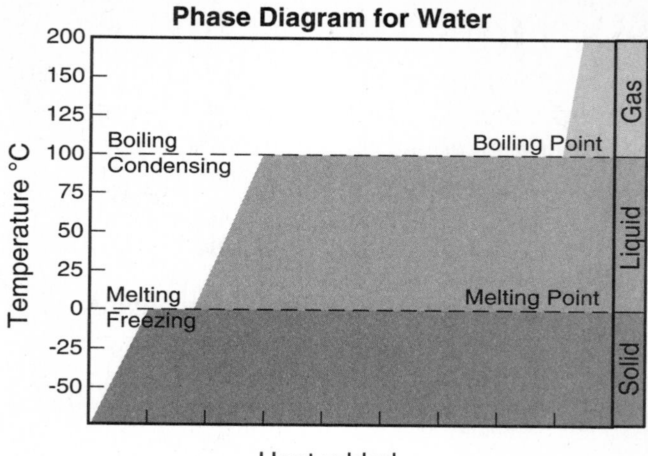

10. How does the melting temperature of solid water compare to the freezing temperature of liquid water?

    (1) The melting temperature is 100° higher than the freezing temperature.
    (2) The melting temperature is slightly higher than the freezing temperature.
    (3) The melting and freezing temperatures are the same.
    (4) The melting temperature is slightly lower than the freezing temperature.
    (5) The two temperatures cannot be compared.

11. The diagram shows that when heat is applied to water to make it boil, the temperature of the water

    (1) stays the same throughout the time the water is boiling
    (2) rises slightly during the time the water is boiling
    (3) decreases during the time the water is boiling
    (4) rises sharply during time the water is boiling
    (5) rises sharply after the water starts boiling and then levels off

Items 12–17 refer to the following information.

A mixture is a combination of substances held together by physical means. Many raw materials are mixtures. Industries separate ingredients by taking advantage of differences in physical properties. The following methods are only some of the ways mixtures can be separated.

extraction—placing a mixture in water or another solvent to dissolve one of its components, filtering off the liquid and then evaporating the solvent, leaving the previously dissolved substance

distillation—vaporizing the mixture and collecting the ingredients as they reliquefy at different temperatures

sorting—selecting the desired ingredients from a fragmented mixture by hand or machine to capture particles according to the size or visual characteristics desired

magnetic separation—using magnets to separate mixtures; the non-magnetic waste drops off, leaving the desired ingredient

gravitation—separating a mixture by the density of the ingredients; the substance with the greatest density settles to the bottom while the substance with the least density remains at the top

12. A series of sieves with different hole sizes is used to separate soil into components of specific grain size. The process of separation is

(1) extraction
(2) distillation
(3) sorting
(4) magnetic separation
(5) gravitation

13. Excess fat is separated from gravy by skimming it off the top with a ladle prior to pouring the gravy into a serving dish. The process of separation is

(1) extraction
(2) distillation
(3) sorting
(4) magnetic separation
(5) gravitation

14. Petroleum is heated to a vapor. As the vapor cools, kerosene, gasoline, and other products are collected at their different condensation points. The process of separation is

(1) extraction
(2) distillation
(3) sorting
(4) magnetic separation
(5) gravitation

15. Vanillin, the vanilla flavor, is separated from the vanilla beans by soaking the beans in alcohol, then filtering the beans from the liquid. The process of separation is

(1) extraction
(2) distillation
(3) sorting
(4) magnetic separation
(5) gravitation

16. Iron ore is separated from waste rock by exposing the mixture to a wide electromagnetic belt. The waste falls away, while the iron ore clings to the belt. This process of separation is

(1) extraction
(2) distillation
(3) sorting
(4) magnetic separation
(5) gravitation

17. A solid substance dissolved in a liquid can be separated out simply by evaporating the liquid. This method is used to obtain

(1) salt from sea water
(2) oxygen from air
(3) carbon dioxide from water
(4) alcohol from water
(5) lubricating oil from petroleum

Items 18–23 refer to the following information.

The lowest temperature at which a substance starts on fire and continues to burn is called its kindling temperature. A flame may provide the heat needed to start a substance burning but is not a necessary factor.

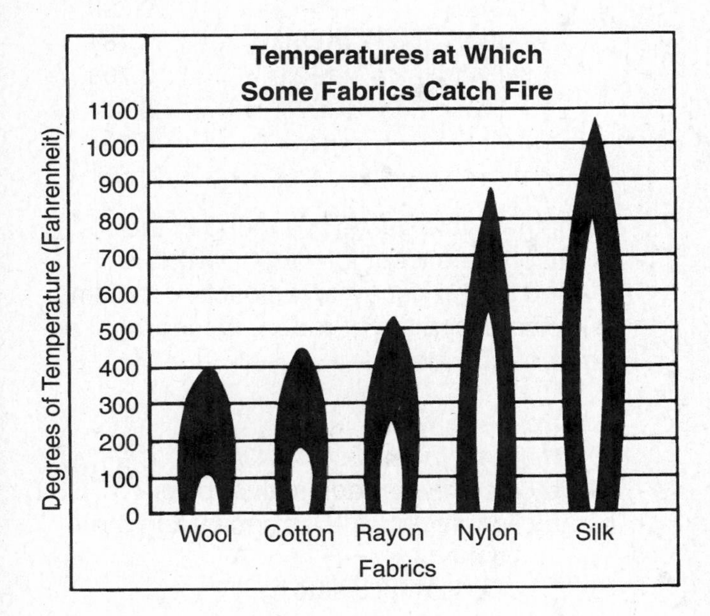

18. A garment made of which of the following fabrics would have the lowest kindling temperature?
    (1) cotton
    (2) nylon
    (3) rayon
    (4) silk
    (5) wool

19. Which fabric has a kindling temperature about half that of silk?
    (1) wool
    (2) cotton
    (3) rayon
    (4) nylon
    (5) silk

20. Which of the following would be the best clothing for a cook?
    (1) a short-sleeved silk shirt
    (2) a long-sleeved nylon shirt
    (3) a cotton shirt with long, flowing sleeves
    (4) a long-sleeved cotton shirt
    (5) a loose, flowing wool shirt

21. In which of the following situations would a fire be least likely?
    (1) continuing to operate a car with a broken fan belt causing the engine to overheat
    (2) leaving a six-pack of carbonated beverages in a car parked all day in the hot sun
    (3) storing rags, dirty from painting chemicals with low kindling temperature, near a furnace
    (4) setting a kerosene heater near cloth upholstery or drapery
    (5) lighting a cigarette while filling a car with gasoline

22. By rubbing matter against matter, the energy of motion is turned to heat. For substances with low kindling temperatures, the rubbing against a rough surface provides enough heat for ignition with the oxygen in the air. Matches are effective igniters because of all of the following factors except
    (1) the head is composed of two different chemicals
    (2) the head is composed of chemicals with low kindling temperatures
    (3) the materials on the scratching surface are rough
    (4) the oxygen in the air is sufficient to support the burning of wood
    (5) the chemicals and wood burn with sufficient heat to ignite many other substances

23. Wool blankets can be used to cover and put out small fires because
    (1) wool has a low kindling temperature
    (2) wool has a high kindling temperature
    (3) wool absorbs water easily
    (4) the blanket prevents the burning item from reacting with the oxygen in the air
    (5) the wool blanket is rough and rubs against the burning item producing sparks

Items 24–25 refer to the following table.

The Noble Gases

Element	Atomic Number	Atomic Weight	Electron Configuration	Oxidation Number	Melting Point, °C	Boiling Point, °C	Density, g/l
Helium	2	4.0026	2	0	–272.2	–268.9	0.177
Neon	10	20.183	2, 8	0	–248.7	–245.9	0.899
Argon	18	39.948	2, 8, 8	0	–189.2	–185.7	1.784
Krypton	36	83.80	2, 8, 18, 8	0	–157	–152.9	3.708
Xenon	54	131.30	2, 8, 18, 18, 8	0	–112	–107.1	5.85
Radon	86	222.	2, 8, 18, 32, 18, 8	0	–71	–61.8	9.73

24. The members of the noble gases family are all rather inert in that they are resistant to forming compounds with other elements. The members are all invisible, odorless, and colorless. A scientist, however, could distinguish between them by all of the following ways except

    (1) weighing equal volumes of the gases
    (2) comparing their natural colors
    (3) determining the boiling points
    (4) determining the densities
    (5) determining the melting points

25. All of the noble gases are found in the atmosphere except for radon, which is formed by the decay of radioactive radium. With the exception of radon, the inert gases are probably obtained

    (1) from sea water
    (2) from mining radioactive rocks
    (3) by electrifying radioactive rocks
    (4) by liquefying air, then separating the liquid mixture
    (5) from industrial waste

Items 26–28 refer to the following passage.

Hydrogen is the lightest of all the gases. Air contains a mixture of different kinds of gases. The heaviest ones sink and are concentrated near the earth's surface. The lightest ones float to the top of the atmosphere and escape into space.

Much hydrogen gas was released into the atmosphere during the early stages of the earth's development. At present, the atmosphere contains 0.001% hydrogen.

26. The passage above implies that the hydrogen originally in the Earth's atmosphere

    (1) escaped into space
    (2) reacted with oxygen to form water
    (3) was present in very small amounts
    (4) is still part of the atmosphere today
    (5) was captured by the gravitational pull of the moon

27. Which statement best explains why there is so little hydrogen in the atmosphere?

    (1) Hydrogen gas unites with oxygen to form water.
    (2) Hydrogen atoms float in the air and escape into outer space.
    (3) Many hydrogen atoms are used in the formation of hydrocarbon fuels.
    (4) Carbohydrates and fats contain carbon, hydrogen, and oxygen.
    (5) The composition of the sun and other glowing stars is primarily hydrogen.

28. Seventy-eight percent of the air at Earth's surface is composed of nitrogen. Therefore, in the mixture called air, nitrogen is

    (1) one of the atmosphere's heavier elements
    (2) very reactive with oxygen
    (3) not reactive with other gases
    (4) not found in the soil
    (5) essential for life

29. Water is known as the universal solvent because it dissolves more substances than any other solvent. In which of the following items is water not used as a solvent?

(1) washing machine
(2) dry-cleaning machine
(3) cup of tea or coffee
(4) soft drink made from a powdered mix
(5) plant roots absorbing fertilizers

30. In a solution, one chemical (the solute) fits into spaces between the particles of the solvent. Which of the following sets of materials is not a solution?

(1) carbonated beverages
(2) a bag of mixed jellybeans
(3) alcohol in a mixed drink
(4) coffee with sugar
(5) fertilizer mixed with water

31. A certain chemical compound is used as a food additive. Which of the following statements best supports a claim that this compound is harmless to people?

(1) The elements making up the compound are all known to be harmless by themselves.
(2) The compound occurs in nature.
(3) Another compound made up of the same elements is known to be harmless.
(4) No one has become ill right after eating a food product containing the compound.
(5) Rats fed large amounts of the compound showed no ill effects over several generations.

32. Entropy is the measure of the amount of disorder in a system. In which of the following is entropy decreasing?

(1) your kitchen gets more cluttered
(2) a fallen tree decays
(3) iron rusts
(4) water freezes
(5) salt dissolves in water

Items 33–35 refer to the following passage.

Constants do not change whereas variables do change. Independent variables change on their own while dependent variables change only in response to changes of another variable. Amedeo Avogadro came to the conclusion that for equal volumes, all gases contained the same number of molecules if the temperature and pressure on the gas were the same. Avogadro estimated this number to be 602,000,000,000,000,000,000,000 (or 6.02×10^{23}). This number came to be called Avogadro's number and is used in many calculations involving gases.

33. Avogadro's number is a constant. Therefore Avogadro's number

(1) will increase in proportion to the amount measured
(2) will decrease as the volume increases
(3) will depend on the dependent variable
(4) will depend on the independent variable
(5) will not change

34. A decrease in the amount of acid added to a specific volume of water will lessen the concentration of the solution. Which statement indicates the relationship of the amount of acid to the concentration of the solution?

(1) The concentration is a constant.
(2) The concentration depends on the amount of acid.
(3) The amount of solution depends on the concentration of the acid.
(4) The solution is concentrated when the acid is concentrated.
(5) There is no relationship between the amount of acid and the concentration.

35. The greater the surface area exposed to the air, the faster water from a boiling solution can evaporate into the air. After a liquid has boiled for a time, the amount of water left in the container

(1) is the independent variable
(2) is a constant
(3) is dependent on the type of solution
(4) is dependent on the surface area exposed
(5) is independent of the amount of water originally in the container

Answers are on page 81.

Unit 4 Physics

Items 1–4 refer to the following graphs.

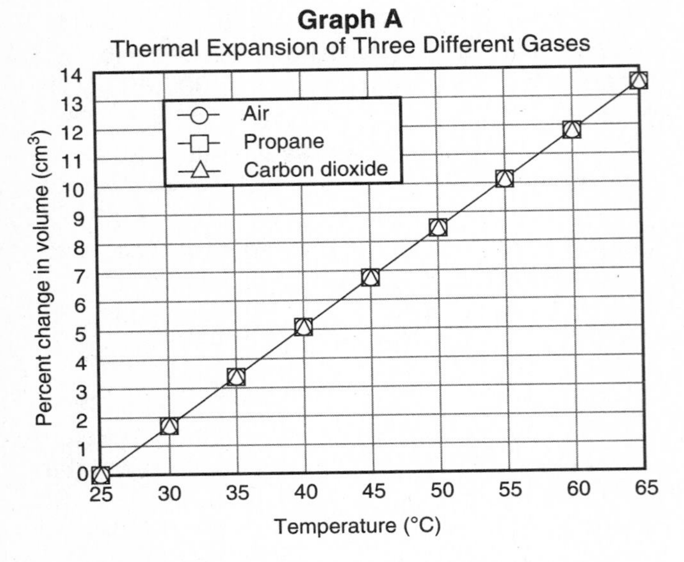

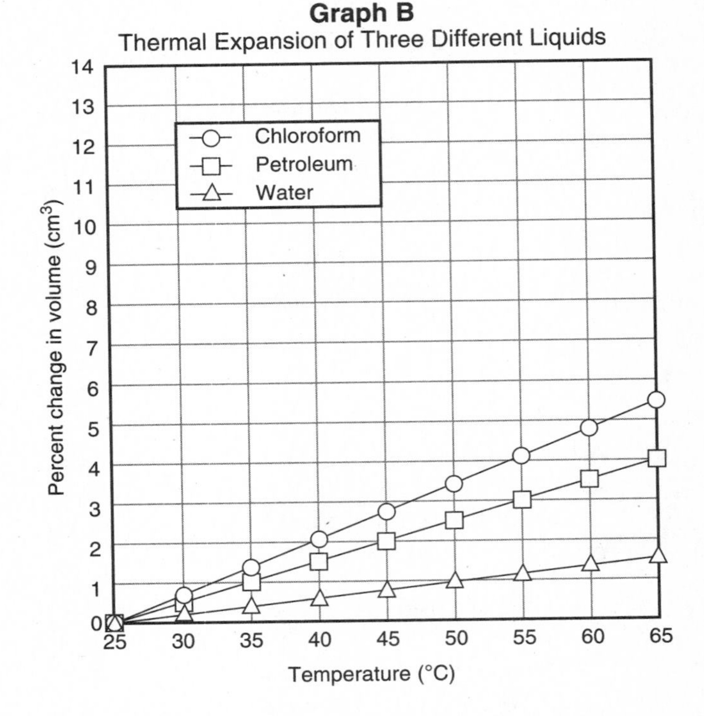

1. Charles's Law states that if the pressure remains constant, equal volumes of different gases will expand the same amount when heated. According to graph A, if a 100 cm^3 sample of gas is heated from 25°C to 55°C, it will expand by
   (1) 3 cm^3
   (2) 4 cm^3
   (3) 5 cm^3
   (4) 10 cm^3
   (5) 50 cm^3

2. Which of the following statements best summarizes the differences between Graph A and Graph B?
   (1) Expansion by heat is dependent on the kind of liquid but independent of the kind of gas.
   (2) Liquids expand more when heated than gases at the same temperature.
   (3) The gases were tested mixed together but the liquids were tested individually.
   (4) For an equal temperature increase, air and water have the same percent volume increase.
   (5) The volume of a gas is dependent on temperature, but the volume of a liquid is independent of temperature.

3. The laws of nature are universal. If a space probe to the planet Venus returns a sample of the gases in Venus's atmosphere to Earth, which of the following results would be expected upon heating the sample?
   (1) The gases would explode violently.
   (2) The gases would contract.
   (3) The gases would not change as Venus is hotter than Earth.
   (4) The gases would expand in conformance to Charles's Law.
   (5) If the gases in the sample are separated, each will expand at a different rate.

4. A 100 cm^3 sample of an unknown liquid at 25°C is heated to 55°C. The volume of the liquid will most likely
   (1) be unchanged
   (2) increase by 3 cm^3
   (3) increase by 10 cm^3
   (4) increase by an unknown amount
   (5) decrease by an unknown amount

See Also

Science Text	Unit 4
Complete Preparation	Unit 4, Physics

Item 5 refers to the following diagram.

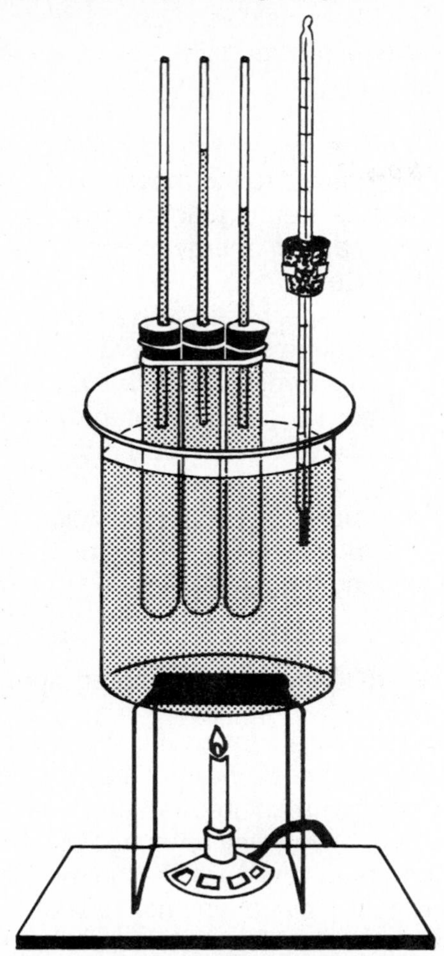

Thermal Expansion of Heated Liquids

5. Observation of the diagram above does not support the conclusion that

   (1) the liquid in the beaker is being heated
   (2) the temperature of the heated liquid is being measured
   (3) the three test tubes contain liquids
   (4) the liquids in the test tubes have expanded in volume
   (5) the air pressure in the thin glass tubes is falling

6. Moving air can act as a force and do work by pushing or pulling calm air to other areas or moving other substances. Which of the following machines does not use moving air to perform its function?

   (1) vacuum sweepers for cleaning
   (2) spray pumps for painting
   (3) sails for sailing boats
   (4) fans for cooling homes
   (5) buzz saws for cutting timber

7. Some machines obtain energy to move their parts by burning a fuel. In order for fuel to burn, it must use oxygen from the air. All of the following machines need air to burn fuel in order to move the parts that make work easier except

   (1) power lawn mowers
   (2) cars and trucks
   (3) sewing machines
   (4) motor boats
   (5) garden tractors

8. When fuels are burned, the energy used causes the resulting gases to expand and push on mechanical moving parts. In which of the following vehicles is the entire vehicle moved by the expanding gases rather than moving parts within the vehicle?

   (1) rocket ships
   (2) diesel engines with pistons
   (3) airplane turbines
   (4) steam engines on trains
   (5) a motorbike with a gasoline engine

9. A lever is a type of simple machine. It transmits a force at one point along its length when a force is applied at another point, while the whole lever pivots at a third point. Which of the following devices makes use of a lever?

   (1) nut and bolt
   (2) wheelbarrow
   (3) ramp
   (4) hammer
   (5) pulley

Items 10–12 refer to the following paragraph.

The plasma state is considered the fourth state of matter. Materials existing as solids, liquids, and gases can enter the plasma state if ionized by very high temperatures in the tens of thousands of degrees Celsius. Much matter on the sun exists in the plasma state.

10. According to the passage above, matter enters the plasma state

    (1) only in the sun
    (2) only from the gas state
    (3) when its pressure is decreased
    (4) when it is heated to extreme temperatures
    (5) when it is bombarded by radiation

11. In which of the following places would one expect to find matter in the plasma state?

    (1) in microwave ovens
    (2) in conventional ovens
    (3) at the equator
    (4) on the moon
    (5) in glowing stars

12. When matter exists in a solid form, it has its own size and shape. All the following substances are solids except

    (1) air in a balloon
    (2) an ice cube
    (3) a roll of paper towels
    (4) a hot dog and bun
    (5) a library book

Item 13 refers to the following diagram.

Steam Turbine Operating a Ship's Propeller

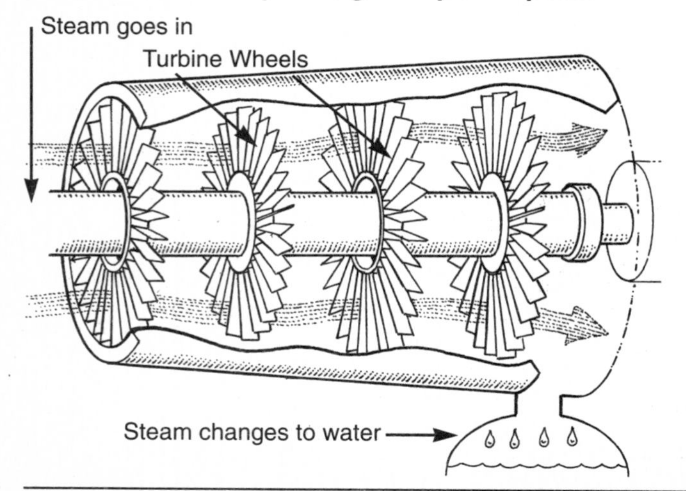

13. Gases have higher energy levels than liquids. The most likely reason water enters the turbine as steam but leaves as liquid water is that

    (1) the energy stored in the steam was transferred to the movable turbine wheels creating motion, and the decrease in energy changed the steam to water
    (2) the turbine wheels were cold, therefore cooling the steam
    (3) the high pressures inside a turbine increased the speed of the water molecules and created motion
    (4) the moving blades cut up the steam molecules into water droplets
    (5) the moving blades acted as fans, cooling the steam

Item 14 refers to the following paragraph.

Photovoltaic cells convert the energy of sunlight into electrical energy. They are fairly expensive to manufacture, and each cell produces a relatively tiny amount of electricity. For these reasons they are not the most cost-efficient way to provide electrical power for large numbers of people. However, using photovoltaic cells produces no harmful by-products and does not deplete our non-renewable petroleum reserves.

14. The information in the paragraph supports which conclusion?

    (1) Photovoltaic cells will never be an important source of electrical power.
    (2) Photovoltaic cells are the best way to produce electricity.
    (3) Photovoltaic cells have some advantages over fossil-fuel-burning power plants.
    (4) Photovoltaic cells are the answer to satisfying our future electric power needs.
    (5) Using photovoltaic cells to produce all our electricity will eliminate pollution.

Items 15–16 refer to the following information.

Uses of Lasers

medical	
• cancer surgery	vaporizes tumor without destroying surrounding tissue
• eye surgery	welds detached retinas and cauterizes bleeding vessels
industry	cuts cloth, metals, and intricate shapes with clean edges
communications	
• fiber optics	no need for expensive copper wires for telephone lines
• photo offset printing	speed in publication
• reads video disks	no need for tapes that deteriorate
commerce	
• reads universal product codes	for accuracy at checkout counters
military	
• death ray	selective rather than general destruction
science	
• accurate measurements	for observing small differences

15. Lasers (Light Amplification by Stimulated Emission of Radiation) were invented in 1954 as an outgrowth of radiation research. At that time lasers had no known use. Which of the following possible comments made at the time of invention would be considered an enlightened position?

    (1) "These rays are too dangerous to ever be of any use."
    (2) "I can't see putting more money into useless light."
    (3) "Only the military will use lasers. We civilians have enough guns."
    (4) "The Death Ray is just science fiction for the movies."
    (5) "I wonder how many ways the future generations will use these interesting rays?"

16. Which of these uses probably requires the most powerful lasers?

    (1) cutting metal in industry
    (2) reading video disks
    (3) reading universal product codes at cash registers
    (4) making scientific measurements
    (5) printing

Items 17–18 refer to the following information.

The destructive effects of an atomic bomb are of five types:

1. flash—bright light that may temporarily or permanently blind one who watches
2. blast—a shock wave that breaks objects into pieces
3. thermal radiation—a release of heat over 3,000°C that vaporizes most substances
4. nuclear radiation—gamma rays and neutrons that destroy living tissue
5. residual radioactivity—small radioactive particles that rise high into the stratosphere, mix with the air, and fall back to Earth's surface. When humans are exposed to it, illness, death, cancer, sterility, and genetic code changes can result.

17. Which of the following effects would a group of individuals who survive a nuclear attack attempt to avoid?

    (1) blast
    (2) flash
    (3) nuclear radiation
    (4) residual radioactivity
    (5) thermal radiation

18. The main effect of a nuclear attack that could cause blindness is

    (1) blast
    (2) flash
    (3) nuclear radiation
    (4) residual radioactivity
    (5) thermal radiation

Items 19–21 refer to the following paragraph.

To produce sound, matter vibrates. In string instruments, the strings vibrate. Percussion instruments are struck to vibrate. Brass instruments have metal valves or a slide to change vibrating air columns. Wind instruments have holes that, when covered, change vibrating air columns. Wind instruments may also have a vibrating reed.

19. To which group of instruments would the bass tuba and trumpet belong?
    (1) brass
    (2) percussion
    (3) strings
    (4) vibrator
    (5) wind

20. To which group of instruments do guitars and banjos belong?
    (1) brass
    (2) folk
    (3) percussion
    (4) strings
    (5) wind

21. What produces sound by vibrating when a gong is struck by a mallet?
    (1) the hands of the player
    (2) the mallet
    (3) the gong
    (4) the frame holding the gong
    (5) the strings suspending the gong

Items 22–25 refer to the following information.

Following are some examples of how Newton's laws of motion affect our daily lives.

A. It takes more gasoline to speed up a heavily loaded truck from 40 m.p.h. to 50 m.p.h. than it does to accelerate a small car the same amount.
B. All cars are equipped with brakes to apply the force necessary to stop.
C. Water squirting from a lawn sprinkler pushes the metal spinner backward, causing it to rotate.
D. When someone steps ashore from a rowboat, the boat slips backward.
E. A bicycle has gears to make going uphill easier.

22. Which example illustrates the following part of Newton's First Law of Motion? A body in motion will stay in motion unless a force acts upon it.
    (1) A
    (2) B
    (3) C
    (4) D
    (5) E

23. Which example illustrates the following part of Newton's Second Law of Motion? The larger the mass, the greater its resistance to a change in velocity.
    (1) A
    (2) B
    (3) C
    (4) D
    (5) E

24. Which example below illustrates the same law of motion as Example B?
    (1) Water flows downhill.
    (2) Earth continues to revolve around the sun.
    (3) A gun recoils when fired.
    (4) Water flows faster when it enters a narrower pipe.
    (5) A heavy object is moved more easily with a lever.

25. Which example illustrates Newton's Third Law of Motion? For every action, there is an equal and opposite reaction.
    (1) A and B
    (2) B and E
    (3) C and D
    (4) A and E
    (5) E only

Items 26–29 refer to the following passage.

Static electricity is a charge on an object that has collected extra electrons by rubbing against a second object. Static electricity is not very useful. When a charge that has built up on an object comes near or touches an object with insufficient electrons, the object discharges the extra electrons. More electrons must then be collected, usually by rubbing them off another object, before there could be another instance of transfer. Electric current is the continual flow of electrons. The flow is transferred by metal wires. Gold, platinum, and silver are excellent metals for wires; however, their cost prohibits their use for the distribution of electricity to households throughout the United States. They are, however, used in extremely sensitive circuits such as computers and in instances where repair is difficult or costly. Lead and aluminum wiring were found to overheat, melt, and cause fires if large amounts of current were used by a household. Iron lacked the flexibility necessary to withstand the bending motion of electrical cords. Copper was found to be the cost-effective choice for general electrical usage and is used extensively throughout the world. It too can overheat, melt, and cause fires but not as quickly as aluminum. As a safety precaution, fuses are required in all circuit installations. Fuses prevent fires by stopping current flow before melting can occur in the wire.

26. Which of the following examples is not related to static electricity?

    (1) the rubbing of molecules in clouds against each other resulting in lightning
    (2) sparks seen and crackling heard when removing a nylon sweater from over a nylon blouse in the dark
    (3) a flashlight shining after exchanging the old batteries for fresh ones
    (4) receiving a shock from the office water fountain situated at the far end of a carpeted hallway
    (5) socks sticking together after being dried in a clothes dryer

27. The most likely metals that would be used in the wiring of the space shuttle and manmade satellites are

    (1) gold and silver
    (2) lead and aluminum
    (3) copper and iron
    (4) lead and copper
    (5) aluminum and copper

28. Consumer usage depends on current rather than static electricity because

    (1) the charge of static electricity is too great
    (2) the rubbing needed to produce static electricity is not cost effective
    (3) the flow of electrons is not continuous with static electricity
    (4) the continuous flow of current electricity is too dangerous
    (5) the static discharge cannot use fuses to prevent fires

29. The persons benefiting most from the improper installation and use of fuses would be

    (1) the original building contractors who want to meet city regulations
    (2) the homeowners who do not wish to risk the lives of family members
    (3) the county budget office that must pay for the maintenance of a fire department
    (4) the labor and repair contractors who repair the damage or demolish the structure and rebuild a new home
    (5) the insurance company that insures a home

Items 30–31 refer to the following information.

Cars and trucks have induction coils to boost the voltage of the electricity from the battery to the spark plugs. Batteries produce direct current of rather low voltage. To increase the voltage, a primary coil and a secondary coil are wound around an iron core. The secondary coil will have a much larger voltage because of the greater number of turns in the wire.

Electricity produced at power plants is alternating current. To push the current through long distances, the voltage must be high. This high voltage is too dangerous to use in households so when it nears the customer, the voltage is decreased to 120 volts. A device to step up or step down the voltage of alternating current is called a transformer. Transformers are designed like induction coils with an iron core surrounding primary and secondary coils.

30. A van or bus would have which of the following devices to supply high voltage current?

(1) stabilizer
(2) induction coil
(3) battery
(4) step up transformer
(5) step down transformer

31. In many foreign countries, the electrical power supplied is 240 volts. When American tourists take electric razors or hairdryers to Europe, they must place which of the following devices between the appliance and the electrical outlet?

(1) battery
(2) dry cell
(3) induction coil
(4) step up transformer
(5) step down transformer

An Induction Coil

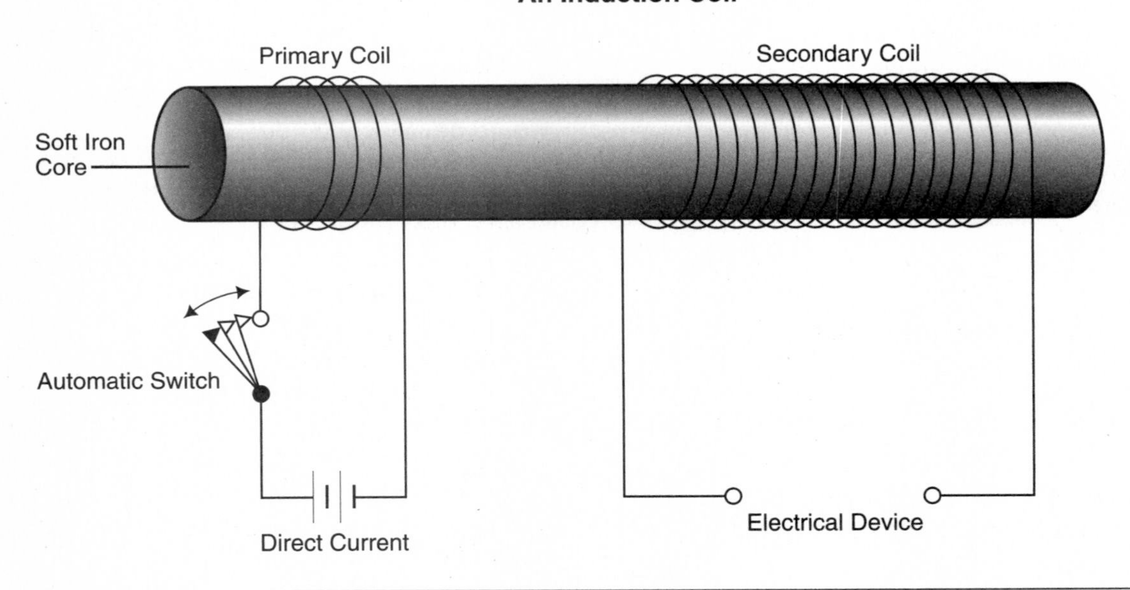

Item 32 refers to the following information.

Mirrors reflect light in real or distorted ways depending on the curvature of their surface. Prisms break light into the visible spectrum, revealing a rainbow of colors. Lenses bend light and can be made to focus light and/or seem to increase or decrease an object's size.

32. Which of the following glass objects would be part of a pair of eyeglasses to increase visual observation at normal distances?

(1) flat mirrors
(2) curved mirrors
(3) triangular prisms
(4) right angle prisms
(5) lenses

Answers are on page 83.

Simulated GED Test A

SCIENCE

Directions

The Science Test consists of multiple-choice questions intended to measure your understanding of general concepts in science. The questions are based on short readings that often include a graph, chart, or diagram. Study the information given, and then answer the questions that follow. Refer to the information as often as necessary in answering the questions.

You should spend no more than 95 minutes answering the 66 questions on Simulated Test A. Work carefully, but do not spend too much time on any one question. Do not skip any items. Make a reasonable guess when you are not sure of an answer. You will not be penalized for incorrect answers.

When time is up, mark the last item you finished. This will tell you whether you can finish the real GED Test in the time allowed. Then complete the test.

Record your answers to the questions on a copy of the answer sheet on page 94. Be sure that all required information is properly recorded on the answer sheet.

To record your answers, mark the numbered space on the answer sheet that corresponds to the answer you choose for each question on the test.

Example: Which of the following is the smallest unit in a living thing?

(1) tissue
(2) organ
(3) cell
(4) muscle
(5) capillary

The correct answer is "cell"; therefore, answer space 3 should be marked on the answer sheet.

When you finish the test, use the Correlation Chart on page 55 to determine whether you are ready to take the real GED Test, and, if not, which skill areas need additional review.

Do not rest the point of your pencil on the answer sheet while you are considering your answer. Make no stray or unnecessary marks. If you change an answer, erase your first mark completely. Mark only one answer space for each question; multiple answers will be scored as incorrect. Do not fold or crease your answer sheet.

Adapted with permission of the American Council on Education.

Directions: Choose the best answer to each item.

Items 1–4 refer to the following passage.

Societies have used branding, tattooing, scar formations, photography, body measurements, genetic configuration, voice printing, and fingerprinting to identify their members. The United States Department of Justice has over 200 million fingerprints on file to assist its Identification Division and local law-enforcement agencies.

If fingers were smooth, they would too easily slide over objects, significantly decreasing the usefulness of hands. The ridges on the bulbs of the fingers are classified by the pattern of arches, loops, and whorls. As no two fingerprints have been found to be exactly alike, fingerprinting is considered an infallible method of identification.

1. Fingerprint identification would be a likely way to investigate the individual identity of all of the following except

   (1) historical skeletal remains
   (2) amnesia victims
   (3) legal alien workers
   (4) unknown dead people
   (5) kidnapping victims

2. The biological purpose of the structures that form fingerprints is

   (1) individual identification
   (2) individual beauty marks
   (3) to assist cleanliness
   (4) to assist in grasping and holding objects
   (5) to assist in retaining skin fluids

3. The height and weight of an individual listed on a driver's license is a means of identification by

   (1) photography
   (2) genetic configurations
   (3) fingerprinting
   (4) branding
   (5) body measurements

4. Identification of sperm by analyzing the DNA of a sample obtained from a rape victim would most likely be by

   (1) fingerprinting
   (2) body measurements
   (3) genetic configurations
   (4) voice printing
   (5) scar formations

5. Heat energy is vital to living organisms. Without heat, all substances would be frozen, preventing any movement. All of the following examples are evidence of the necessity for heat to activate life processes except

   (1) normal human body temperature is 98.6°F
   (2) seeds become dormant under freezing temperatures
   (3) lizards can run in the afternoon sun but become sluggish during cool nights
   (4) bears hibernate in winters
   (5) seals and walruses swim in cold ocean water

Items 6–9 refer to the following information.

Chemists use the following symbols to denote the conditions and products of chemical interactions:

yields $\rightarrow$
equilibrium $\rightleftharpoons$
gas is formed $\uparrow$
solid is formed $\downarrow$
heat is needed for the reaction $\xrightarrow{\Delta}$
light is needed for the reaction $\xrightarrow{light}$

Reactants are indicated on the left side and products are indicated on the right side of an equation.

$$6CO_2 + 6H_2O \xrightarrow[chlorophyll]{light} C_6H_{12}O_6 + 6O_2$$

The above equation means that the two reactants, carbon dioxide (CO_2) and water (H_2O) in the presence of light energy in cells that have chlorophyll, can be changed into sugar ($C_6H_{12}O_6$) and oxygen gas (O_2). The numbers preceding a substance indicate how many molecules are needed. For example, $6CO_2$ means six CO_2 molecules are needed for the interaction to take place.

6. The equation $2HgO \xrightarrow{\Delta} 2Hg + O_2\uparrow$ shows the breakdown of mercury oxide (HgO) to mercury (Hg) and oxygen(O). All of the following statements about this equation are true except

   (1) heat is needed for the reaction to take place
   (2) the mercury and oxygen were separated
   (3) oxygen is given off as a gas
   (4) the reactants and products reached equilibrium
   (5) two molecules of mercury oxide were necessary for the reaction to take place

7. A chemist using symbols to represent reactions between chemicals is most like

   (1) a secretary using shorthand
   (2) a cook using ingredients
   (3) a swimmer using goggles
   (4) an artist using paints
   (5) a doctor using gloves

8. Which of the following statements is known to be true because of a symbol used in the equation below?

   $$Zn + 2HCl \longrightarrow ZnCl_2 + H_2\uparrow$$

   (1) A gas is produced.
   (2) Zinc is a metal.
   (3) Hydrochloric acid is a liquid.
   (4) Not all the zinc will be dissolved.
   (5) Heat is needed for the reaction.

9. Equations are likely to be found in all the following places except

   (1) the chalkboard of a chemistry classroom
   (2) the experimental notes kept by a research chemist
   (3) a pharmacist's handbook on chemical mixtures and compounds
   (4) a scientific journal for chemical engineers
   (5) a daily newspaper article about the discovery of a new compound to treat AIDS

Items 10–11 refer to the following illustration.

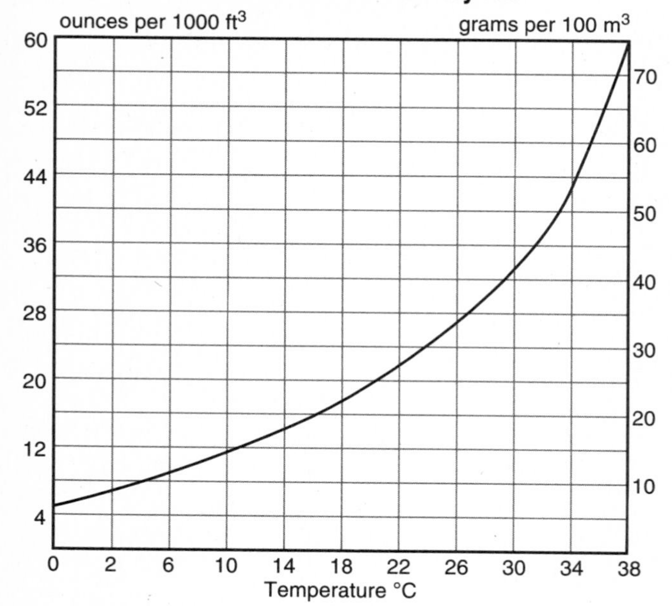

10. Approximately how much water, in grams per 100 m^3, can air hold at a temperature of 10°C?

(1) 8
(2) 11
(3) 14
(4) 20
(5) 48

11. Which statement enables the above graph to be used internationally?

(1) The information cannot be used since the graph's author is not listed.
(2) The information presented is accurate globally.
(3) The information is presented in both English and metric units.
(4) The information is about scientific facts.
(5) The information cannot be used since the water in air differs from one locality to another.

12. Relative humidity is the percent of water in the air for a particular air temperature. The daily relative humidity report would be important to all of the following persons except

(1) a TV weather forecaster
(2) an asthmatic child with severe breathing problems
(3) a caterer for an outdoor barbecue
(4) a cosmetologist styling the hair of the bride for a wedding ceremony
(5) a deep-sea diver investigating sunken treasure

Items 13–17 refer to the following passage.

Natural selection is the way nature chooses which organisms survive. Chance mutations occur in response to chemicals or certain kinds of energy in the electromagnetic spectrum. If the mutant is better adapted to the environment, it thrives. If not, it dies out or becomes rare.

Humans have used artificial selection to produce plants and animals with desirable characteristics. Many of these domesticated plants and animals can no longer survive in the wild. Their survival depends on the maintenance of an artificial environment and the desires of people.

People select certain desired traits such as color, beauty, or scent (as in roses). Other traits that are bred artificially include uniqueness (as in the neck plumage of the prized Jacobin pigeon), size (as in miniature horses), meat quality or milk yield (as in cattle), or resistance to disease (as in fungus-resistant tomatoes). The traits usually are selected for convenience, pleasure, or financial gain of individuals. In this way, humans act as agents of evolution through artificial selection.

Individual specimens with the desired traits are crossbred. The hybrid offspring are then inbred to preserve and fix the desirable characteristics and eliminate unfavorable characteristics from the stock.

A pure breed is formed when there is not any mixture of other genes over many generations. The American Kennel Club recognizes 121 breeds of purebred dogs. When ancestors of a pure breed are known and registered by a breed club, the dog is said to have a pedigree.

13. Some people argue that scientific tampering with plants and animals by artificial selection will do more harm than good. Which statement best supports this argument?

(1) The dog is now human's best friend.
(2) Many hybrids and pure breeds can no longer survive in the wild.
(3) There are now 121 breeds of purebred dogs.
(4) Humans are agents of plant and animal evolution.
(5) Inbreeding fixes desirable characteristics.

14. The dog as an example of artificial selection is supported by all but which statement?

(1) Dogs are domestic animals.
(2) There are 121 recognized breeds of dogs.
(3) Breeders register dogs to obtain a pedigree.
(4) Humans have been the primary agents in dog evolution.
(5) The dog is one of nature's survivors.

15. Which of the following results of artificial selection is least beneficial to humanity?

(1) Financial gain is increased considerably by producing better plants and animals.
(2) Specific varieties and breeds are obtained that would not have occurred in nature.
(3) Humans must maintain the artificial environments necessary for the survival of the selected breeds and varieties.
(4) Humans are able to control plant and animal reproduction for humans' pleasure.
(5) Many new kinds of plants and animals are produced.

16. Breeding the hybrid offspring to fix traits is called

(1) hybridization
(2) artificial selection
(3) inbreeding
(4) pedigree
(5) purebreed

17. A farmer imported several fine long wool Tomney sheep from Australia to breed with his Debouittet sheep, which produce under the difficult conditions on the New Mexico range, in hopes of increasing the value of the flock's wool. This is an example of

(1) purebreeds
(2) inbreeding
(3) crossbreeding
(4) hybridization
(5) fertilization

Items 18–20 refer to the following information.

Domestic Members of the Species *Brassica oleracea*

Common Name	Part of Plant Eaten
cabbage	terminal bud
cauliflower	flowers
kohlrabi	stem
Brussels sprouts	lateral buds
broccoli	flowers & stem
kale	leaves

Form

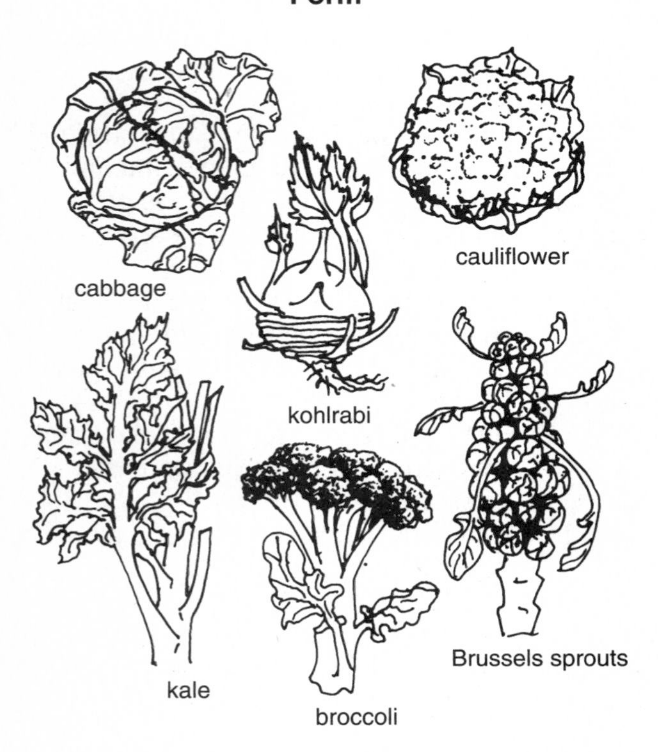

18. Which of the following influences is most likely responsible for the diversity and heightened development of the specialized food storage organs in the above plants shown?

(1) natural selection
(2) selective intervention by humans
(3) wind dispersal
(4) ceremonial and religious usage
(5) climatic adaptations

19. The plant part not specialized for food storage in the species Brassica oleracea is the

(1) leaf
(2) stem
(3) bud
(4) root
(5) flower

20. The members of the species Brassica oleracea listed can no longer survive in the wild. This fact supports all but which of the following statements?

(1) Total dependence is a consequence of domestication.
(2) Many edible wild species can propagate themselves.
(3) Survival of the fittest rules in nature.
(4) The end result of intense selection is the demand for artificial conditions.
(5) Domestic plants and animals are bound together with humans in adaptive co-evolution.

Items 21–23 refer to the following illustrations.

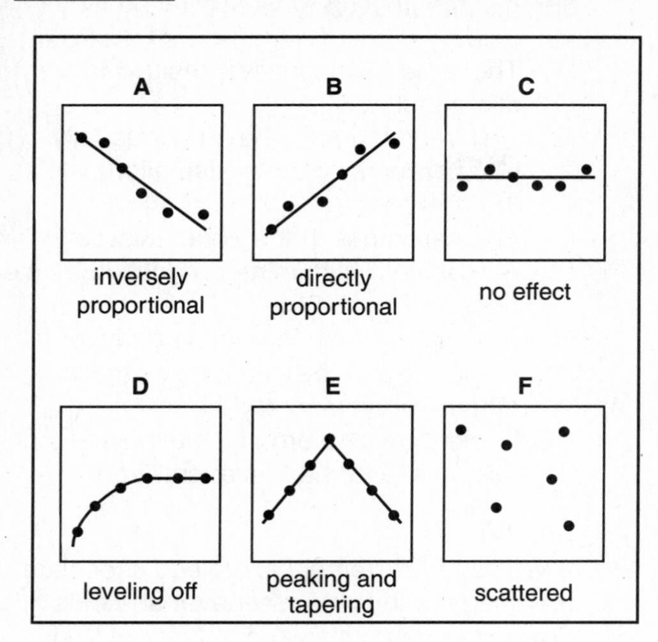

21. A company wishes to show that their nuclear power reactors have not in any way endangered the lives of people in areas surrounding the nuclear power plants. Which graph would best reflect this contention?

    (1) A
    (2) B
    (3) C
    (4) D
    (5) E

22. Which of the graph types would be used by an environmental lobbyist in attempting to show the amount of animal life remaining as an oil spill invades a bay?

    (1) A
    (2) B
    (3) C
    (4) E
    (5) F

23. After being heated, a cold solution at 0°C reached its boiling point at 87.5°C. Further heating did not increase the liquid's temperature beyond that point. Which graph would best illustrate the heating of the liquid?

    (1) A
    (2) B
    (3) C
    (4) D
    (5) E

Items 24-28 refer to the following information.

Chemicals called allergens occur in many household substances. Allergens can be found in mold spores, pollen, dust, and dandruff or hair of pets. Many foods, especially chocolate, eggs, cow's milk, wheat, and shellfish also contain allergens. People who are sensitive to certain allergens respond to the chemicals by producing histamine which causes capillaries to enlarge, mucous glands to secrete, and smooth muscle to tighten. The enlarged capillaries cause hives, headaches, and other tissue swelling. The nasal drip of allergic rhinitis and the phlegm produced in the bronchial tubes of asthmatics are excess gland secretions. The tightening of the smooth muscle may cause the upset of the entire gastrointestinal tract.

To diagnose a particular allergy, needles are treated with chemicals from suspect substances. The needles are used to place the substances under the skin. Records are kept of the location of each substance injected. If a red patch appears where the skin is pricked, an allergy to that substance is indicated.

There is no cure for allergies. Avoidance of the known allergen is the best treatment. For some allergy sufferers, this simply means getting rid of a pet or not eating shellfish. For others, avoidance may be almost impossible, as in allergies to ragweed pollen or house dust. In severe attack cases, doctors called allergists may prescribe antihistamine drugs or attempt to desensitize the person by introducing small but constant levels of allergens into the bloodstream.

24. Which statement best indicates the basic cause of all allergies?

(1) Shellfish contain chemicals that produce one of the most common food allergies.
(2) Avoiding contact with house dust is almost impossible.
(3) Certain people's bodies react to chemicals called allergens by producing histamines.
(4) Certain people's bodies react to chocolate by tightening the smooth muscles of the gastrointestinal tract.
(5) Allergens can be found almost everywhere.

25. Which statement best explains why a nasal drip accompanies hay fever?

(1) The nose is unusually sensitive to allergens.
(2) Children who have hay fever usually have parents who are also allergic to the allergen.
(3) The capillaries in the nose enlarge in response to histamine production by the body.
(4) Histamine produced by the body in response to allergens causes mucous glands to oversecrete.
(5) Avoidance of allergens is almost impossible for hay fever sufferers.

26. A woman is extremely allergic to chocolate, breaking out in hives whenever she eats it. For over a year, this woman has not had symptoms of the allergy. The most likely reason is that the woman is

(1) avoiding foods with chocolate
(2) not producing histamines in the body when eating chocolate
(3) eating only milk chocolate
(4) no longer taking antihistamine drugs
(5) no longer worried about an allergic attack

27. Antihistamine drugs are

(1) the only treatment for allergies
(2) used to desensitize persons with allergens
(3) used to avoid severe symptoms
(4) used to alleviate symptoms in severe attacks
(5) used to avoid contact with allergens

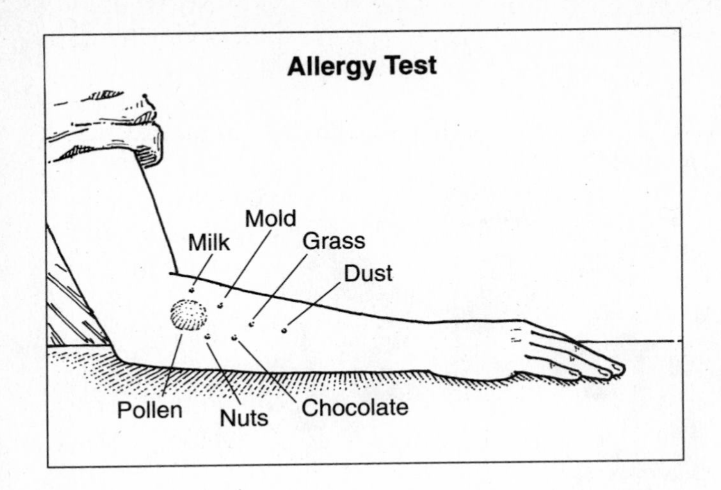

28. From the diagram, which statement can be supported by evidence for the individual being tested?

   (1) The individual is only allergic to the pollen and not to other substances being tested.
   (2) The individual is allergic to dust and nuts.
   (3) The family of the individual is not allergic to milk or mold.
   (4) The parents of the individual are also allergic to the kind of pollen injected.
   (5) The individual is allergic to all types of pollen.

29. Machines for providing exercise have become important to people interested in body building. Muscle tissue can only be formed from protein. Which of the following statements best describes how muscle tissue is increased by the use of body building machines?

   (1) Request for increased energy to move machine parts stimulates the body to produce muscle tissue from proteins.
   (2) Many muscle builders eat large quantities of meat.
   (3) The decrease of body fat by burning it while using machines increases the proportion of muscle.
   (4) The force needed to operate a machine requires increased oxygen from deep breathing.
   (5) Body building machines require the users to burn high-energy foods thus increasing their appetite.

Items 30–31 refer to the following map.

Global Malaria Distribution

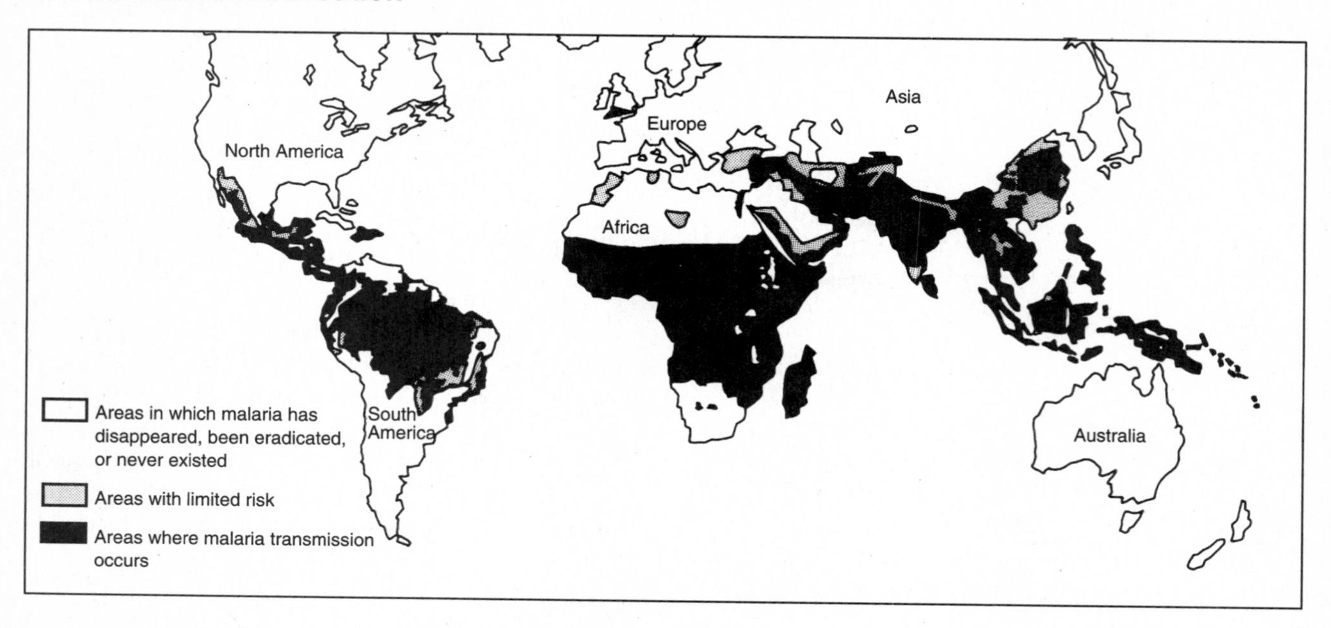

30. Which of the following is a conclusion that can be drawn based on the map above?

    (1) In most cases of malaria diagnosed in the United States, the individual contracted the disease outside the U.S. border.
    (2) Effective programs have eradicated malaria in Southeast Asia.
    (3) Australia is likely to become an area of limited risk for malaria in the near future.
    (4) The cold and temperate areas of the world present the greatest risk in contracting malaria.
    (5) Malaria is not a serious threat to most of Earth's population.

31. Which of the following facts is of least importance for an American traveling to Central Africa?

    (1) Regardless of the type of protection taken, it is still quite possible to contract malaria.
    (2) Appropriate drugs that offer protection from malaria increase the odds of not contracting the disease.
    (3) Even after a person completes a protection program, a fever can indicate onset of malaria.
    (4) The risk of contracting malaria in Central Africa is high.
    (5) In northern and southern Africa, the incidence of malaria is limited.

Item 32 refers to the following passage.

Tetanus is often a fatal human disease. Sometimes called lockjaw, tetanus is caused by Clostridium tetani bacteria that live in the soil in large numbers. A wound that gets dirt or soil in it is likely to contain these bacteria.

Tetanus bacteria multiply only in the absence of oxygen. Thus, surface wounds exposed to the air rarely allow the bacteria to multiply. Deep wounds, however, are ideal for the development of the disease.

32. A condition that would greatly increase the probable development of tetanus would be

    (1) a bullet wound during a store robbery
    (2) a puncture wound while digging in the garden
    (3) a floor burn while playing basketball
    (4) a hairline fracture to the skull during an auto accident
    (5) a mosquito bite

Item 33 refers to the following passage.

All body cells, fluids, and gases are chemicals. These chemicals interact in the processes that maintain life. When an organism is not functioning normally, often the cause can be determined by testing cells, gases, or body fluids with other chemicals called indicators. These indicators change color or form an identifiable new chemical.

33. Which of the following individuals is performing a medical test that does not involve the use of chemicals in the testing?

    (1) a diabetic using a glucose test strip for sugar in the blood
    (2) a young woman using a urine-based home pregnancy test
    (3) a possibly intoxicated driver performing a breathalyzer test after an accident
    (4) a woman manually examining her breasts for possible tumors
    (5) a technician testing the acid-pH level of a patient's stomach fluids

Item 34 refers to the following information.

A doctor instructed a diabetic patient to use a glucose strip indicator twice a day in the following manner.

1. Place one drop of blood on the strip, and wait 2 minutes.
2. Wipe off the blood, and compare the top and bottom colors to those listed below.
3. Act in accordance to the action suggested by the matching colors

Top / Bottom		Action
light tan / pale blue	=	eat a sugar substance immediately
tan / blue	=	normal, no action required
light green / blue	=	restrict sugar and carbohydrate intake
dark green / bright blue	=	call the doctor for an appointment
black / dark blue	=	go directly to the hospital

34. A blood sample was tested and the strip indicator turned tan over blue. The patient should

    (1) call the doctor and arrange for an appointment
    (2) follow normal eating patterns
    (3) eat a substance containing sugar immediately
    (4) severely limit foods containing sugar
    (5) be driven to the nearest hospital

35. If organic means must contain carbon, inorganic means non-living, astro means among the stars, bio means life, and engineering means science of machines and structures, then the study of the chemicals in body cells, fluids, and gases would most likely be called

    (1) chemical engineering
    (2) astrochemistry
    (3) biochemistry
    (4) organic chemistry
    (5) inorganic chemistry

36. As you watch distant fireworks, you notice that you hear the sound of the explosions a few seconds after you see the flashes of light. This observation is evidence that

    (1) sound energy and light energy travel at the same speeds
    (2) sound energy travels faster than light energy
    (3) light energy travels faster than sound energy
    (4) light energy travels in a straight line but sound energy does not
    (5) ears process information faster than eyes

Items 37–42 refer to the following passage.

Modern American homes use large amounts of energy. Much of the required energy is purchased from suppliers of electrical power. Electricity in itself is not the desired end product. People do not use electricity directly, but they use it after it has been changed into other energy forms.

Generators produce electrical current by using running water, or heat (from burning fuels, or nuclear fission) to turn turbines. The turbines provide the motion to move a coil of wire through a magnetic field. The magnetic field produces the electricity.

Appliances change electricity to the desired energy: from motors to motion, resistors to heat or light, and magnetic vibration to sound. Excessive usage, causing overheating, can damage appliances.

Transformers increase and decrease the voltage for transport and safety. Switches enable the owner to decide usage; while fuses prevent excessive current that could overheat the wires and ignite fires. Insulators prevent access to currents that could cause death if touched. Most Americans learn to use electricity safely and wisely.

37. Which of the following practices would contribute most to the electrical safety of young children in a household?

(1) keeping the lights on all night
(2) placing protectors in unused electrical outlets
(3) using electric blankets thus keeping the air temperature lower
(4) using a humidifier to increase the percentage of water in the air
(5) installing a gas water heater in place of an electric one

38. Which of the following situations does not demonstrate that electricity can be changed into other energy forms?

(1) A gas stove has an automatic pilot controlled by electricity.
(2) A fan cools the air when connected to electricity.
(3) At the hydroelectric plant, a turbine shaft moves a wire coil through a magnetic field generating electricity.
(4) Sound comes from a radio plugged into a circuit outlet.
(5) The temperature of the air in a house remains constant as the thermostat controls the electrical heating system.

39. An electric motor turns electrical energy into mechanical motion. Which of the following appliances best illustrates an application of the above statement?

(1) fluorescent lighting
(2) radio and TV
(3) blenders and mixers
(4) refrigerators and freezers
(5) curling irons and toasters

40. Which of the following is not likely to result from long continuous use of an electrical appliance?

(1) wearing out the essential parts
(2) an increased electrical bill
(3) overheating and fire
(4) more frequent maintenance such as lubrication
(5) an increase in the longevity of the appliance

41. Which of the following devices is used to control the time for appliance usage?

    (1) fuse
    (2) switch
    (3) transformer
    (4) insulator
    (5) wires

42. Which of the following relationships is most like that of an insulator to the current in the wires?

    (1) a cover to the book in a library
    (2) a rug to the floor in a house
    (3) a shade to the lamp on a desk
    (4) a fur coat to the mannequin in a showcase
    (5) a double cage to a lion in a zoo

Item 43 refers to the following information.

Heat flows, or moves from place to place, in three ways.

Conduction—Heat energy is the vibration of atoms or molecules in matter. Heat flows by conduction when rapidly vibrating molecules bump into their neighbors and make them vibrate faster.

Convection—In a fluid, which can be a liquid, a gas, or a flowing mixture, heat often flows by convection. When part of the fluid is heated, it becomes less dense than the surrounding fluid and rises. This movement of the fluid carries heat energy to other parts of the fluid.

Radiation—Heat can be transferred to an object by radiant energy. Radiant energy from the sun passes through space and the atmosphere. When it strikes an object, it makes that object's molecules vibrate faster.

43. You place a pot of water on the heating element of an electric stove. As the water goes from cold to boiling, heat is flowing by

    (1) conduction only
    (2) convection only
    (3) conduction and convection
    (4) radiation and convection
    (5) radiation and conduction

44. The force with which one object hits another depends on the direction of motion, mass, and speed of both objects. Understanding this information is of great value to all the persons below except

    (1) a mathematician calculating the weight of several objects
    (2) a pilot maneuvering take-off into the wind
    (3) a boxer in a championship match
    (4) a driver in expressway traffic
    (5) a football tackle hoping to sack the quarterback

Items 45–47 refer to the following diagram.

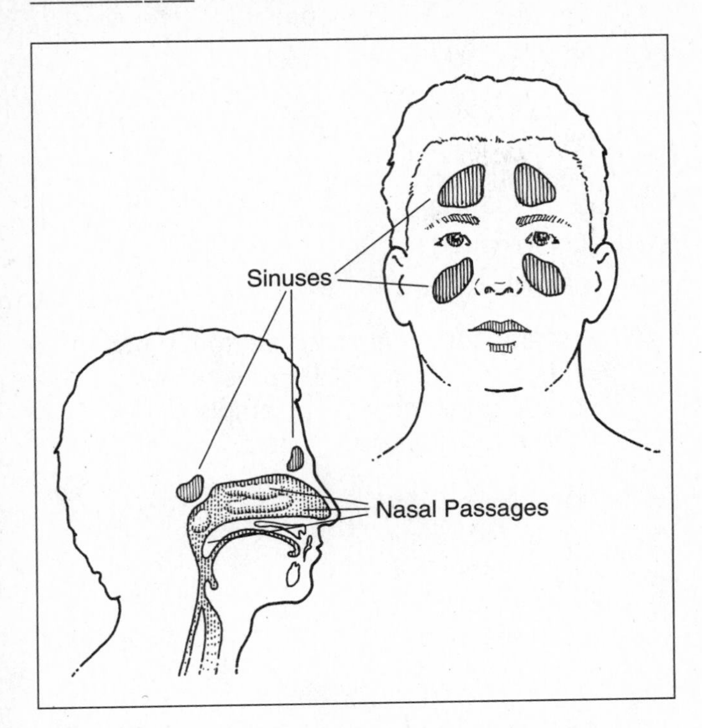

45. When sinuses become partly clogged and inflamed, the individual is least likely to experience

    (1) the pain of pressure on the eyes resulting in squinting
    (2) frontal headaches
    (3) pain in the upper jaw and misinterpret the cause as toothache
    (4) dripping from the nostrils
    (5) pressure on the back of the skull resulting in diminished vision

46. Sinus pain results when tubes leading away from the sinuses swell, trapping mucous. Which of the following actions would be most effective in relieving sinus pain?

    (1) extracting some maxillary teeth
    (2) plugging the nasal cavities
    (3) stroking the frontal bone
    (4) lying down with the eyes closed
    (5) sniffing a vapor that causes the mucous to flow

47. Which of the doctors listed below would probably hear the most complaints regarding sinus pain?

    (1) surgeon
    (2) cardiologist
    (3) family physician
    (4) podiatrist
    (5) dermatologist

Items 48–51 refer to the following information.

Instruments are essential to scientific investigation because the human body has (1) no detectors for certain forms of energy, (2) limited sensors for specific sizes, distances, and vibrations, and (3) limited ability to accurately measure what is observed. Instruments are designed for specific tasks that increase the range and accuracy of observation. The following instruments are examples of many used by scientists to obtain information that is otherwise unavailable.

microscope—used to make small objects appear larger
oscilloscope—used to see variations of electrical current as wavy lines on a screen
periscope—used to get a view from a level above that of the eyes
spectroscope—used to identify elements by analyzing the arrangement and amount of colored light emitted by an element
telescope—used to make distant objects appear nearer and larger

48. Binoculars used at a football game or opera are a type of

(1) microscope
(2) oscilloscope
(3) periscope
(4) spectroscope
(5) telescope

49. The screen of a heart monitor displaying the electrical activity of a patient in intensive care is a type of

(1) microscope
(2) oscilloscope
(3) periscope
(4) spectroscope
(5) telescope

50. A throat culture is obtained from a patient and grown. To verify the presence of streptococci bacteria, a sample of the culture will be viewed with a

(1) microscope
(2) oscilloscope
(3) periscope
(4) spectroscope
(5) telescope

51. An astronomer publishes an article in which the composition of the sun is purported to be mostly hydrogen and helium. The instrument probably used by the astronomer to reach this conclusion is a

(1) microscope
(2) oscilloscope
(3) periscope
(4) spectroscope
(5) telescope

Item 52 refers to the following graph.

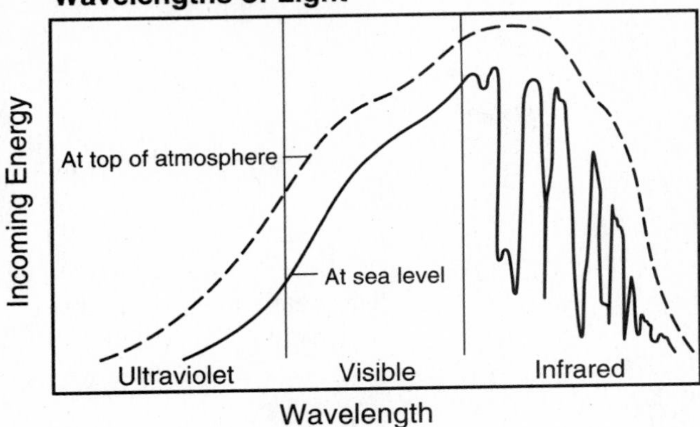

52. Compared to plants growing on Earth, plants grown in a space satellite orbiting Earth would receive

    (1) less ultraviolet light energy
    (2) less infrared light energy
    (3) less of all light energy wavelengths
    (4) more of all light energy wavelengths
    (5) more of some infrared light wavelengths only

Items 53–54 refer to the following information.

Ten percent of the inquiries to the U.S. Poison Control Centers concern plants. The greatest number of calls concern children under three years of age. Often these children have access only to plants in the home. Older children tend to ingest plants and mushrooms in the yard or playground; whereas teenagers and adults become intoxicated by foraging for wild edible plants. Sometimes intoxications associated with presumably harmless plants are due to chemical contaminations by pesticides, herbicides, and fertilizers.

53. Which of the following actions would have the greatest impact on the reduction of plant poisonings of children under three years of age?

    (1) the publication of books about plant poisonings
    (2) eliminating mushrooms from the family diet
    (3) eliminating or preventing access to houseplants until the child is over three years of age
    (4) keeping houseplants free from infestation by using pesticides
    (5) increasing the fertilization of houseplants to insure healthy plants

54. The information regarding plant poisonings would most likely appear in all of the following types of books except

    (1) home child care
    (2) wilderness survival
    (3) medical handbook on toxic substances
    (4) Red Cross First-Aid Manual
    (5) plant propagation text

Items 55–56 refer to the following illustration.

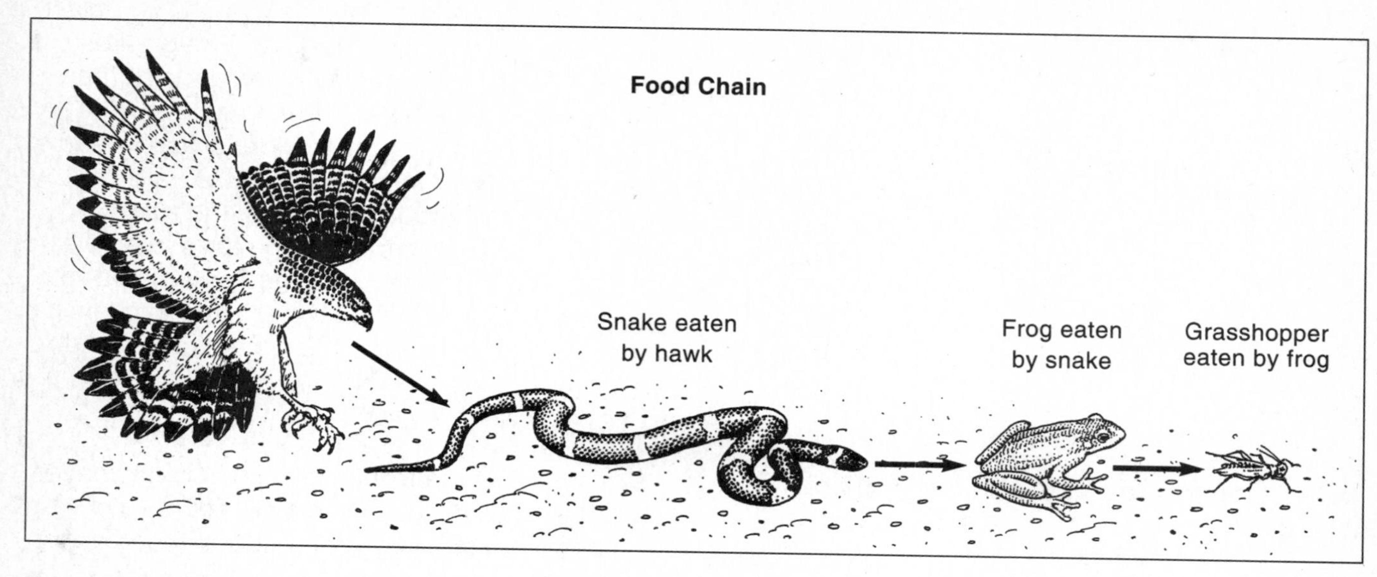

55. The most noticeable characteristic of the order of the animals in this food chain is

    (1) means of locomotion
    (2) length
    (3) size
    (4) speed
    (5) eye type

56. The major advantage of being at the top of a food chain is that

    (1) all those below are easier to catch
    (2) the top animal can be seen more easily
    (3) there are more choices in what the animal can eat
    (4) the top animal can fly
    (5) the bottom animal can hide more easily

57. When a tree is cut down, the top, middle, and bottom all land on the ground at the same time. According to this observation, which of the following inferences must be correct?

    (1) The middle must control the ends to move at the same speed.
    (2) The large mass of the trunk moves the bottom faster.
    (3) The many branches equal the weight of the trunk.
    (4) The top must move faster than the middle or bottom.
    (5) The top, middle, and bottom all move at the same speed.

Items 58–61 refer to the following passage.

Radon, always present in the air, is an invisible, odorless, and tasteless radioactive gas. Granite and shale rocks can contain uranium which undergoes a natural breakdown to form radon gas. If a house is built over these rocks, the radon gas can seep in through drains, cracks in concrete, and pores in hollow block walls. The gas then accumulates, particularly in basements and crawl spaces.

In the United States, at least 5,000 deaths a year can be attributed to radon gas. The risk of developing lung cancer from exposure to radon depends on the time exposed and the concentration level of the gas. Unless the home is tested, the occupants cannot know if they are at increased risk for lung cancer. The U.S. Environmental Protection Agency (EPA) is concerned about the link of radon gas to lung cancer and is sponsoring studies to identify areas of the country that are likely to have indoor radon problems. Preliminary studies indicate that in radon-contaminated homes, other factors also contribute to the risk of lung cancer. Smoking, amount of time spent in the home, sleeping in the basement, and the number of years lived in the home all increase cancer risk.

Test kits have been developed to enable individuals to assess the radon levels of their home or workplace. The EPA offers free information about radon risk levels and permanent, cost-effective solutions to radon problems. Radon test kits are available in most hardware stores and pharmacies.

58. Radon gas causes all of the following except

(1) an increased risk of lung cancer
(2) enough concern by federal agencies to sponsor research to identify areas at risk
(3) a change in the taste of water or the smell of the air
(4) at least 5,000 deaths per year in the U.S.
(5) studies to determine its effects

59. Accurate information about specific radon risk levels and cost-effective solutions to the problem could best be obtained from

(1) the local hardware store
(2) the pharmacist at the pharmacy
(3) a neighbor whose house has been tested
(4) the Environmental Protection Agency of the federal government
(5) a radon control company representative

60. Which factor has not been identified as increasing the cancer risk from a house having an elevated radon level?

(1) sleeping in the basement
(2) the number of years of residence in the house
(3) eating saturated fats with high cholesterol levels
(4) the amount of time spent in the home per day
(5) smoking

61. Which action would not decrease the risk of lung cancer?

(1) patching cracks in the basement floors and walls
(2) changing jobs after the workplace tested positive for radon
(3) ceasing to smoke cigarettes
(4) building a house over shale or granite rocks
(5) moving from a basement apartment to one on the first floor

Items 62–66 refer to the following information.

The five descriptions below refer to special types of mixtures.

aerosol — a suspension of a liquid in a gas
gel — a suspension of a liquid in a solid
tincture — a substance dissolved in alcohol
amalgam — the mixture of another metal with mercury
biocolloid — particles suspended in a liquid within a living organism

62. Blood has two main parts, the liquid plasma and solid cells, along with undissolved particles, that are suspended in the plasma. Blood is

(1) an aerosol
(2) a gel
(3) a tincture
(4) an amalgam
(5) a biocolloid

63. Vanilla, iodine, and many cough syrups are sold as tinctures. Which of the following substances must they contain?

(1) water
(2) mercury
(3) alcohol
(4) gelatin
(5) sugar

64. Alloys are mixtures of two or more metals. Which of the types of mixtures below could form an alloy?

(1) aerosol
(2) gel
(3) tincture
(4) amalgam
(5) biocolloid

65. Hair spray is an example of

(1) an aerosol
(2) a gel
(3) a tincture
(4) an amalgam
(5) a biocolloid

66. A chef dissolved a powdered mixture in hot water. Pieces of canned fruit were added. Upon cooling, the water and fruits were held firmly in the molded dessert. The chef's dessert was in part

(1) an aerosol
(2) a gel
(3) a tincture
(4) an amalgam
(5) a biocolloid

Answers are on page 85.

Analysis of Performance: Science Simulated Test A

Name: ______________________ **Class:** ______________ **Date:** ________

The chart below will help you determine your strengths and weaknesses in reading comprehension and in the science content areas of biology, Earth science, chemistry, and physics.

Directions

Circle the number of each item that you answered correctly on the Simulated GED Test A. Count the number of items you answered correctly in each column. Write the amount in the total correct space of each column. (For example, if you answered 24 biology items correctly, place the number 24 in the blank before out of 33.) Complete his process for the remaining columns.

Count the number of items you answered correctly in each row. Write that amount in the total correct space of each row. (For example, in the comprehension row, write the number correct in the blank before out of 14.) Complete this process for the remaining rows.

Test A Analysis of Performance Chart

Item Types:	Biology (Unit 1)	Earth Science (Unit 2)	Chemistry (Unit 3)	Physics (Unit 4)	Total Correct
Comprehension	1, 2, 16, **19,** 24, 27, 63	**10,** 58	**8,** 33, **34,** 35	41	_____ out of 14
Analysis	5, 14, **20,** 25, **28,** 29, **30,** **45, 55**	**11,** 60	**6, 7,** 64	36, 38, 40, **52,** 57	_____ out of 19
Application	3, 4, 17, 26, 32, 49, 50, 53, 54, 62	**21, 22,** 48, 51, 61	**23,** 65, 66	39, 42, 43	_____ out of 21
Evaluation	13, 15, **18,** **31, 46, 47, 56**	12, 59	**9**	37, 44	_____ out of 12
Total Correct	_____ out of 33	_____ out of 11	_____ out of 11	_____ out of 11	Total correct: ____out of 66 1–50 = You need more review 51–66 = Congratulations! You are ready!

(**Boldface** items indicate those questions with a graphic stimulus.)

If you answered fewer than 51 questions correctly, determine which areas are hardest for you. Go back to the *Steck-Vaughn GED Science* book and review the content in those specific areas.

In the parentheses under the heading, the units tell you where you can find specific instruction about that area of science in the *Steck-Vaughn GED Science* book. Also refer to the chart on page 3.

Simulated GED Test B

SCIENCE

Directions

The Science Test consists of multiple-choice questions intended to measure your understanding of general concepts in science. The questions are based on short readings that often include a graph, chart, or diagram. Study the information given, and then answer the questions that follow. Refer to the information as often as necessary in answering the questions.

You should spend no more than 95 minutes answering the 66 questions on Simulated Test B. Work carefully, but do not spend too much time on any one question. Do not skip any items. Make a reasonable guess when you are not sure of an answer. You will not be penalized for incorrect answers.

When time is up, mark the last item you finished. This will tell you whether you can finish the real GED Test in the time allowed. Then complete the test.

Record your answers to the questions on a copy of the answer sheet on page 94. Be sure that all required information is properly recorded on the answer sheet.

To record your answers, mark the numbered space on the answer sheet that corresponds to the answer you choose for each question on the test.

Example: Which of the following is the smallest unit in a living thing?

(1) tissue
(2) organ
(3) cell
(4) muscle
(5) capillary

① ② ● ④ ⑤

The correct answer is "cell"; therefore, answer space 3 should be marked on the answer sheet.

When you finish the test, use the Correlation Chart on page 75 to determine whether you are ready to take the real GED Test, and, if not, which skill areas need additional review.

Do not rest the point of your pencil on the answer sheet while you are considering your answer. Make no stray or unnecessary marks. If you change an answer, erase your first mark completely. Mark only one answer space for each question; multiple answers will be scored as incorrect. Do not fold or crease your answer sheet.

Adapted with permission of the American Council on Education.

Directions: Choose the best answer to each item.

Items 1–2 refer to the following information.

For some people, a single bee sting is a serious matter. Within a number of minutes of being stung, an individual may start wheezing, experience difficulty swallowing, become pale, and eventually collapse. The person should remain calm, minimize activity, and seek medical attention.

1. Which of the following describes the attitude that people hypersensitive to bee stings should take when stung by a bee?

(1) excited
(2) concerned
(3) defeated
(4) harsh
(5) indifferent

2. Which of the following relationships is most like a bee sting to a person who is hypersensitive to bee venom?

(1) rattlesnake bite to a dog
(2) cat bite to a mouse
(3) fish bite to the fish bait
(4) heart attack to a human
(5) accident to a driver

Items 3–4 are based on the following diagram.

Levels of Complexity for Lifeforms

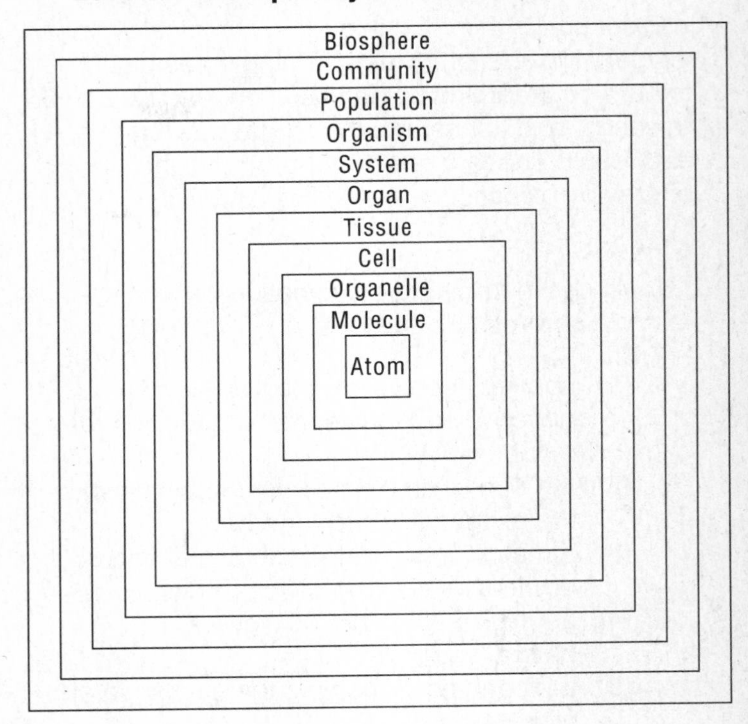

3. All of the statements are supported by the diagram except which one?

(1) Life can be organized into a series of levels by complexity.
(2) Cells are made up of organelles.
(3) The basic building blocks of life are atoms.
(4) The living aspect of the biosphere consists of communities.
(5) Molecular development is dependent on cellular formation.

4. Which of the following series is correctly ordered from least to most complex?

(1) atoms, cells, communities, organs
(2) molecules, cells, tissues, atoms
(3) organs, tissues, cells, molecules
(4) organisms, populations, communities, biosphere
(5) biosphere, communities, population, organisms

Items 5–6 are based on the following information.

Oxidation is the combining of a chemical with oxygen. If oxidation is fast, burning takes place. In burning, heat and sometimes light energy are released, and ashes or exhaust gases are the end result. The fast release of heat or light energy by oxidation is called fire.

5. All of the following combinations with oxygen result in fire except

    (1) spark plugs igniting the mixture of gasoline and oxygen in the cylinders of a car engine
    (2) the iron of an old bicycle combining with the oxygen in air to form rust
    (3) a match used to start the combining of dry leaves with the oxygen in air
    (4) a furnace combining oxygen with natural gas
    (5) a jet engine combining the oxygen in air with kerosene

6. In the body, food is used as fuel. Oxygen is taken in through the lungs. The blood transports both oxygen and food to the cells where burning takes place. Which body process is evidence that burning is taking place in the body?

    (1) growth in size
    (2) reproduction
    (3) maintenance of body temperature
    (4) repair of injured cells
    (5) circulation of blood

7. Motors turn electricity into motion; whereas engines turn chemical energy into motion by burning a fuel. Which of the following means of transport is most likely to contain a motor rather than an engine?

    (1) airplane
    (2) semi-truck
    (3) submarine
    (4) oil tanker
    (5) subway train

8. In a highly industrialized technological society, power to run machines is an essential commodity. Which of the following legislative actions by Congress would have the least effect on increasing the supply of power?

    (1) providing money to build dams and hydroelectric plants
    (2) opening off-shore and Alaskan oil fields for drilling
    (3) providing federal funds for research in the use of nuclear fuels
    (4) establishing land grant colleges to increase agricultural productivity
    (5) providing rural electrification grants to provide all localities in the U.S. with electricity

Items 9–12 refer to the following passage.

Butterflies and moths are insects that belong to the order Lepidoptera. Both develop through stages of complete metamorphosis. In the butterfly, the pupa is encased in a hardened covering called a chrysalis; while the moth larva spins a strong, but soft, cocoon that surrounds the pupa.

When the adult emerges, more differences distinguish the moth from the butterfly. Moths generally fly at night; while butterflies are active in the daytime. When at rest, butterflies hold their wings vertically; while moths at rest keep their wings horizontal. Moths have a stout abdomen and feathery antennae; while the more slender butterflies have knobbed antennae.

Lepidoptera aid in cross-pollination as they visit flowers to obtain nectar. The greatest economic value of Lepidoptera is the production of silk by the silkworm, which is not a worm at all but the larva of a small domesticated moth. Silkworms are grown commercially in China and India.

Other larva are notorious for their damage to plants. The gypsy moth is a menace to trees in the northeastern states. The apple worm, cabbage worm, tomato horn worm, and corn-ear worm are all larvae of the Lepidoptera. All larvae are voracious eaters. The adults spend most of their stage seeking mates, with egg laying as the end result.

9. A flower seed company would find butterflies most useful in assisting with

(1) butterfly reproduction
(2) chrysalis spinning
(3) apple production
(4) flower pollination
(5) worm destruction

10. Humans profit most from Lepidoptera through

(1) spider web spinning
(2) exotic wing collecting
(3) elm tree pollination
(4) nectar and honey production
(5) silk production

11. An insect with beautiful mosaic-patterned wings lands on a leaf and lays her eggs. These later hatch as caterpillars which eat ravenously, grow rapidly, and molt several times. Finally, each caterpillar spins a cocoon and rests. The insect is a

(1) worm
(2) Hymenoptera
(3) butterfly
(4) moth
(5) either a moth or a butterfly

12. On what basis would a New York suburban community consider spraying the town to eliminate Lepidoptera?

(1) the proliferation of the butterfly weed
(2) the possible destruction of oak trees
(3) the competition with bees for nectar
(4) the commercial competition of nylon and silk
(5) the failure of a Lepidopterist's convention

Items 13–14 are based on the following illustration.

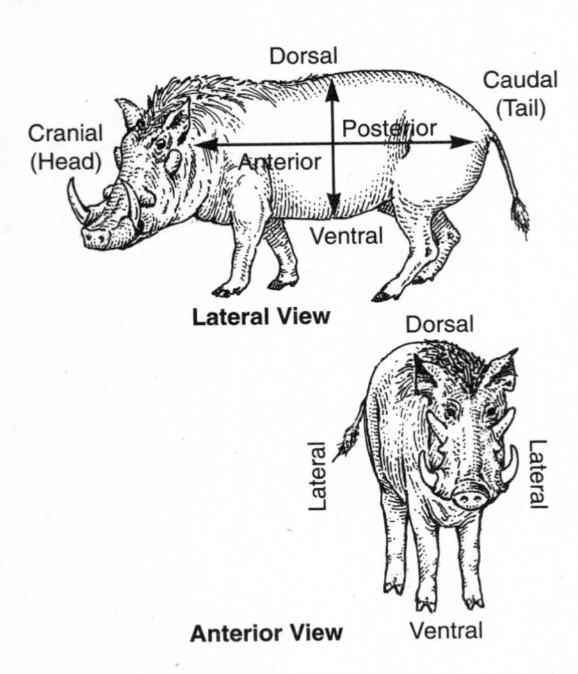

13. The fin extending above the backbone of a fish is most likely called
    (1) a dorsal fin
    (2) a posterior fin
    (3) a caudal fin
    (4) a ventral fin
    (5) a lateral fin

14. A view from behind an object is
    (1) a lateral view
    (2) an anterior view
    (3) a posterior view
    (4) to the right of the ventral view
    (5) a cranial view

Items 15–16 refer to the following list.

A Partial List of "Lilies" That Are Not Lilies

Lily	Botanical Name
African blue lily	Agapanthus africanus
Angel's lily	Ismene
Australian Spear lily	Doryanthes palmeri
Aztec lily	Sprekelia formosissima
Barbados lily	Amaryllis belladonna
Basket lily	Hymenocallis calathina
Blood lily	Haemanthus coccineus
Blue lily of the Nile	Agapanthus umbellatus
Bugle lily	Watsonia
Butterfly lily	Hedychium coronarium

15. The partial list indicates that
    (1) lilies are usually classified alphabetically
    (2) the botanical names of lilies can be written in English
    (3) very large lilies grow in Australia
    (4) the plant known as the Barbados lily is not botanically a lily
    (5) the Liliaceae family has more species than most botanical families

16. The most likely intent for constructing the list was to
    (1) assign the botanical names for common lilies
    (2) identify the proper botanical classification for plants mistakenly called "lilies"
    (3) indicate the extremely large number of species in the lily family
    (4) identify common names for members of the lily family
    (5) classify all nonlily plants

Items 17–20 refer to the following passage.

Many products that society uses are natural substances from plants, animals, and minerals. Synthetics begin with natural substances, are processed by humans, and result in products that do not occur in nature.

To stay alive, the human body must synthesize very complex chemical substances. When the body malfunctions or is hindered by injury or disease, complex chemicals are often required to return the body to a healthy state. Sulfa drugs and aspirin are examples of useful modern synthetic medicines. When the body is under stress or injury, prostaglandins are produced to increase blood flow to the area, raise body temperature, and intensify pain. Aspirin (acetylsalicylic acid) blocks the production of prostaglandins. Increased temperature speeds the chemical activity needed for healing, and increased blood flow provides more cells and chemicals to the area to fight disease or repair damage. Pain alerts the conscious body to possible danger and the need for treatment.

17. Americans swallow nineteen billion aspirin each year. Which of the following is not achieved by using aspirin?

    (1) Injuries are repaired and diseases are cured.
    (2) Pain and suffering are diminished.
    (3) Fever is lowered to more normal levels.
    (4) The production of prostaglandins is blocked.
    (5) The swelling and inflammation caused by increased blood flow are reduced.

18. Which of the following products is a synthetic?

    (1) orange juice
    (2) cotton cloth
    (3) gold chain
    (4) plastic cups
    (5) leather

19. The reason modern societies are able to prolong life is that

    (1) they have higher I.Q.'s
    (2) they have healthier babies
    (3) they are able to synthesize complex substances
    (4) they are literate and able to treat themselves
    (5) they are able to eliminate injury and the causes of stress with painkilling drugs

20. All of the following are results of society's ability to produce complex synthetics except the use of

    (1) insecticides and fungicides to prevent many plant diseases
    (2) nonbiodegradable products that form garbage that cannot be reduced to natural products
    (3) drugs to relieve the pain and stress of modern life
    (4) fibers from petroleum to make nylon, orlon, and polyester clothing
    (5) a wide variety of foods to obtain essential vitamins and minerals by eating

Items 21–25 refer to the following information.

An arrangement of equipment with a swinging rod and attached object is called a pendulum. Sometimes pendulums are paired to work together. An experimenter attached balls of equal mass and size to string of the same thickness in the manner below. Each pendulum was pulled back exactly 30 centimeters and left to swing. The experimenter counted the number of complete swings in one minute. Each ball was swung three times with the following results.

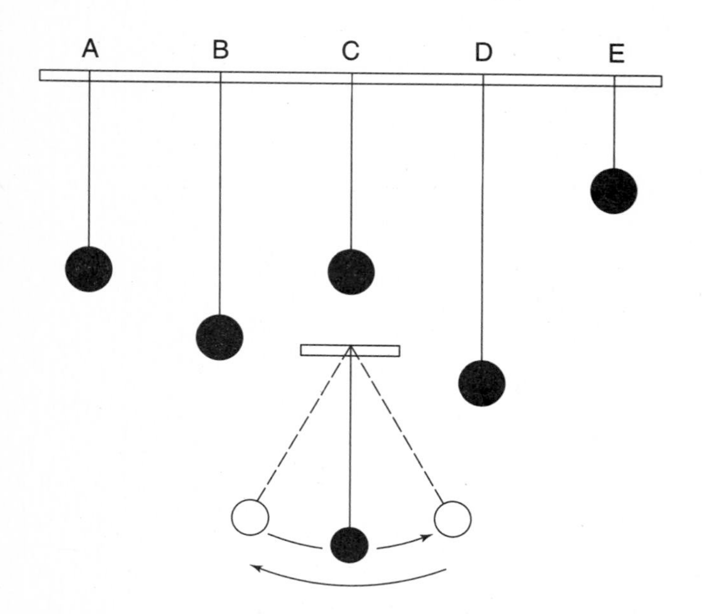

Number of Swings Per Minute of 5 Pendulums

Pendulum	Trial 1	Trial 2	Trial 3	Average
A	22	22	22	22.0
B	10	11	10	10.3
C	21	22	22	21.7
D	8	8	8	8.0
E	44	43	43	43.3

21. The factors that were controlled by the experimenter to remain constant during the experiment were

(1) time only
(2) time and size only
(3) time, size, and mass only
(4) time, size, mass, string thickness, and distance of pull
(5) time, size, mass, distance of pull, and string length

22. The hypothesis used by the experimenter to design the experiment was that the number of swings was dependent on

(1) the size of the ball
(2) the mass of the ball
(3) the length of the string
(4) the distance of pull
(5) all of the above

23. Which of the following changes would best control the experiment if the experimenter wished to know if the mass of a ball affected the number of swings per minute?

(1) only string lengths to be the same
(2) string length the same and vary the mass of the balls
(3) vary the distance of pull and control string length
(4) vary the size of the balls only
(5) vary the mass and the size of the balls

24. The result of the experiment would lead one to conclude that the number of swings per unit time is

(1) constant
(2) dependent on the mass attached to the string
(3) dependent on the length of the string
(4) dependent on the thickness of the string
(5) independent of any factor

25. Pendulums are part of all of the following examples except

(1) playground rubber tire swings
(2) grandfather clocks
(3) wrecking balls used in building demolition
(4) flying trapeze equipment in the circus
(5) roller coasters at amusement parks

26. Which of the following situations does not demonstrate the use of a gas in forming or supporting the structure of an object?

(1) the raising of yeast breads and cakes
(2) a flat tire causing difficulty in the movement of a car
(3) a balloon bursting when stuck with a pin
(4) the need for living things to breathe air
(5) whipped cream losing its shape and returning to a liquid

Items 27–29 refer to the following information.

Keys are helpful for identifying organisms based on their characteristics. The key below is part of a larger key to the trees of a certain mountain range. In a key of this type, you start by choosing one of the two statements at the first level (A or AA). Then, under your choice, you choose one of the two statements at the next level (B or BB). You continue in this way until you arrive at the organism's name.

A. Leaves needle-like (conifers)
 B. Cones hang down
  C. Needles in bunches
   D. Needles in bunches of five—Pinus lambertiana
   DD. Needles in bunches of three
    E. Cones greater than 8 inches long—Pinus coulteri
    EE. Cones less than 5 inches long—Pinus ponderosa
  CC. Needles single—Pseudotsuga menziesii
 BB. Cones point upwards—Abies concolor
AA. Leaves flattened and broad, not needle-like (broadleaf trees)

27. According to the key, Pinus coulteri is distinguished from Pinus ponderosa by

(1) length of needles
(2) length of cones
(3) how the cones grow on the tree
(4) the number of needles in a bunch
(5) type of leaf

28. A tree from the mountain range has cones that hang down and needles in bunches of five. It is

(1) an Abies concolor
(2) a Pinus coulteri
(3) a Pinus ponderosa
(4) a Pinus lambertiana
(5) a broadleaf tree

29. The person most likely to use this key in his or her work would be a

(1) forester
(2) fisher
(3) lumber mill operator
(4) woodworker
(5) rancher

Item 30 refers to the following illustration.

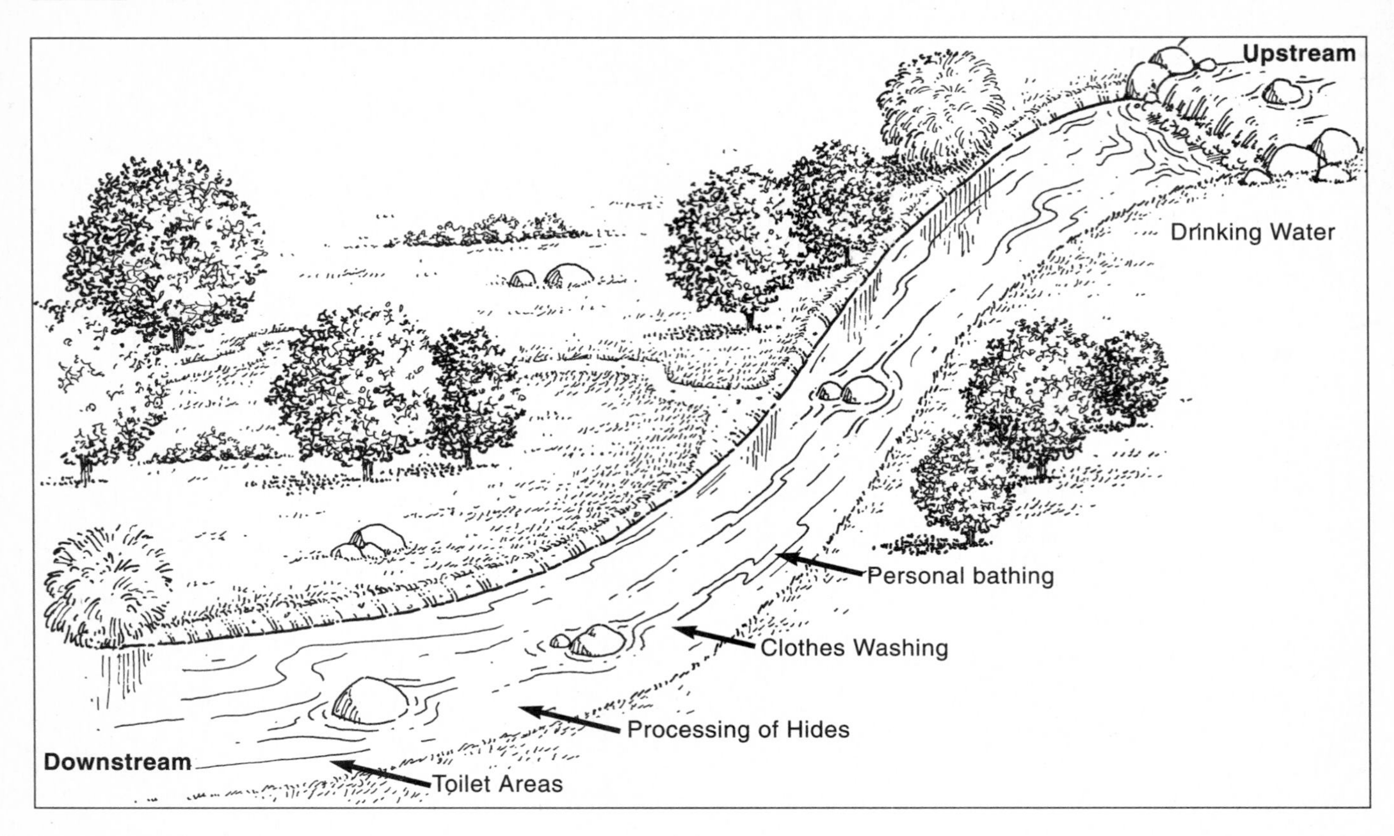

30. A primitive village had strong taboos (restrictions) about the use of a nearby stream. Which change would most likely have the greatest negative effect on the village's water supply?

    (1) construction of a government school
    (2) a medical outpost established in the village
    (3) a dye business assigned to the stream between the processing of hides area and the toilet area
    (4) another tribe building a village close by upstream
    (5) another tribe building a village close by downstream

Items 31–33 refer to the following passage.

Soil contains rocks and minerals of many kinds and sizes. The sizes vary from boulders, large stones, and gravel to gritty sand, silt, and powdery clay. Silt is composed of particles smaller than sand but larger than clay.

In addition to rocks and minerals, waste products and the remains of dead plants and animals are found in soil. These materials, called humus, increase the fertility of the soil. Topsoil contains much humus; while subsoil usually contains very little.

Geologists classify soils into three basic types: clay, sand, and loam. Soil type is determined by the preponderance of particle size and type. Most soils are mixtures of differing proportions of two or more materials.

Clay soil is mostly clay with a little sand and silt. Although clay soil is soft and powdery when finely ground, it holds much water and becomes sticky when wet. Clay soil can become extremely hard in drought conditions when the particles bind firmly together.

Sandy soil consists mostly of sand with a little clay and silt. Because of the larger particle size of sand grains, water easily drains from sandy soil. Sandy soil is less fertile than other types, since many of the chemicals in any humus present dissolve in water and are flushed from soil.

Loam soil combines sand, clay, and silt. It drains water better than clays, contains more humus than sand, and is considered excellent for most crops. However, some crops grow best in sandy or clay soil.

31. According to the passage above, the subsoil is not as fertile as the topsoil because it

(1) is further underground
(2) is not able to hold as much water
(3) contains less humus
(4) has larger sized particles
(5) has very little clay in proportion to sand

32. Before choosing to produce a particular crop, a farmer would be wise to

(1) use as much fertilizer as possible
(2) decrease the amount of sand in the soil
(3) increase the amount of clay in the soil
(4) know only the character of the soil available
(5) know both the soil characteristics and the soil needs of various crops

33. A backyard gardener would like to include a small watermelon patch in a garden with clay/loam soil. According to the seed package, watermelon plants require higher than average drainage. The gardener would probably wish to

(1) increase the proportion of sand
(2) increase the proportion of clay
(3) decrease the proportion of sand
(4) significantly increase the amount of humus
(5) take off several inches of topsoil

Item 34 refers to the following illustration.

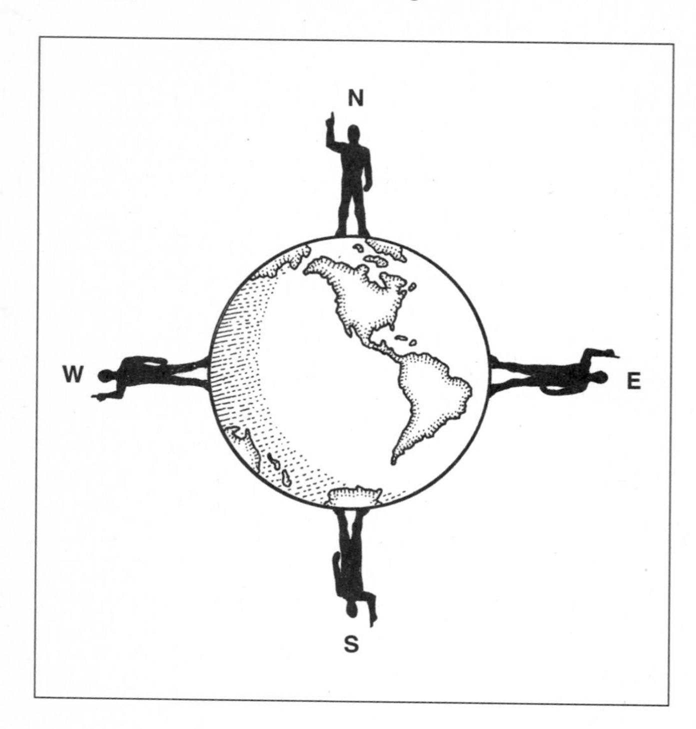

34. Each individual in the diagram has his or her right arm pointing upward. Which of the following statements defines up?

(1) Up means toward the North Pole.
(2) Down means toward the South Pole.
(3) Up means east or west.
(4) Up means away from Earth.
(5) Down means toward the equator.

Items 35–40 refer to the following information.

A warp is the hole in the matrix of space around any object. The greater the mass of the object, the greater the warp. Whether or not another object in the warp can escape depends on the speed of the object attempting to leave and the size of the warp. Gravity is the word scientists often use to describe the effect on an object caught in another object's warp.

Earth's Warps

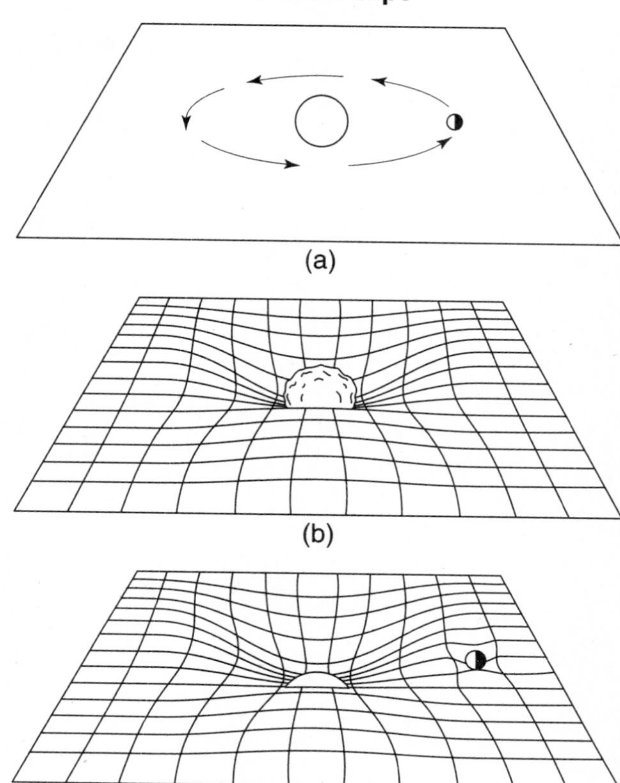

35. The reason people do not fall off Earth into space is that

(1) air pressure pushes objects against Earth
(2) Earth's spin on its axis sucks objects to itself
(3) Earth's magnetism attracts all objects
(4) objects are caught in the warp created by the mass of Earth
(5) people live only on the up side of Earth

36. The speed of the moon is sufficient to keep it from falling further into Earth's warp but insufficient to pull out of Earth's warp. Therefore, the moon constantly orbits Earth. A similar explanation would explain why Earth

(1) rotates on its axis
(2) is tilted on its axis
(3) orbits the sun
(4) has day and night
(5) has seasons

37. Energy traveling at 186,000 miles per second is sufficient to escape the warps in our solar system. Which of the following persons need not be concerned with the size of Earth's warp?

(1) a physicist in charge of the kind and amount of fuel needed for a solar probe
(2) an astronaut in charge of the flight maneuvers for an exploratory trip to the moon
(3) a computer analyst in charge of the space shuttle's orbital path and re-entry
(4) an astronomer on Earth recording radio signals from radiating stars in the Andromeda Galaxy
(5) an aerospace engineer in charge of putting a communication satellite into orbit

38. The planet Jupiter is 318 times more massive than Earth. To escape Jupiter's warp would require

(1) the same force as to escape Earth
(2) a greater force than to escape Earth
(3) a lesser force than to escape Earth
(4) finding a hole in the warp
(5) an object with greater mass than Jupiter

39. Tides along an ocean's shores are the effect of the overlapping warps of Earth, moon, and sun. The most likely reason why the ocean water, unlike the rocks and soil, moves in relation to the warps is that the ocean is

(1) liquid
(2) cold
(3) vast
(4) deep
(5) blue

40. If the weight of an individual depends on the gravity (warp force) of the object on which the person is being weighed, and if the moon has only one-sixth of the gravity that Earth's warp causes, then how much would a person weighing 180 lb. on Earth weigh on the moon?

(1) 1,090 lb.
(2) 360 lb.
(3) 150 lb.
(4) 30 lb.
(5) 15 lb.

Items 41–42 refer to the following information.

Sunscreens contain chemicals that filter out the ultraviolet radiation that causes skin to burn. Recently scientists have determined that skin cancer is directly related to previous amounts of sunburn.

Sunscreens for Differing Skin Types

*SPF of Sunscreen	Skin Type	Frequency of Burns
SPF 15	fair, never tans	always
SPF 10–15	medium, faint tan	occasionally
SPF 4–10	medium, tans well	occasionally
SPF 2–4	dark, tans well	never

*SPF = Sun Protection Factor found on product labels

41. Which of the following facts is most important in determining what SPF product is appropriate for a particular individual?

(1) family cancer history
(2) prevalence of cancer in the area
(3) light eyes and hair color
(4) amount of time spent in the sun
(5) skin type and history of sunburns

42. Which statement is the most relevant to a decision to use a sunscreen?

(1) The current fashion colors look best on persons with dark or tanned skin.
(2) The melanin in dark skin filters out much of the ultraviolet rays that cause burning.
(3) Exposure to the sun is now known to be the primary cause of skin cancer.
(4) Sand and water reflect and intensify the ultraviolet rays that cause burning.
(5) Paraminobenzoic acid is the active ingredient in most sunscreens.

Items 43–44 refer to the following graph.

Number of Legs of Selected Animal Groups

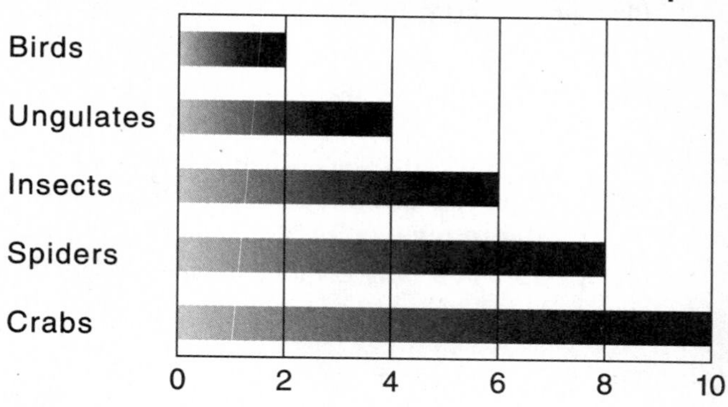

43. From the graph it can be stated that

(1) all winged animals have six legs
(2) sea creatures usually have more legs than land creatures
(3) a characteristic that distinguishes a spider from an insect is the number of legs
(4) crabs have more complex bodies than insects
(5) if an animal is not an ungulate (hoofed animal), it cannot have four legs

44. If a zoologist working in the Amazon jungle discovered a new species of animal that has two legs, the zoologist could immediately

(1) identify it as a bird
(2) identify it as an ungulate
(3) know that it is not an ungulate
(4) know that it will not bite
(5) know that it cannot live in water

45. Classification systems are based on observable differences between members of a group. Which factors would not be suitable as differences to distinguish between kinds of land animals?

    (1) presence, location, and type of hair
    (2) number of toes and type of toenails
    (3) age and total number of offspring
    (4) ear shape and location
    (5) presence and length of tail

46. The presence of water is essential for life. Which of the following body reactions does not indicate the need for increased water intake?

    (1) dry mouth with little saliva
    (2) parched and cracked lips
    (3) tear production
    (4) dry skin with cracks at elbows and heels
    (5) thirst

47. Fungi survive best in dark, warm, moist areas. Athlete's foot is a fungus that often thrives in the warm darkness between the toes of persons whose feet perspire easily. Which of the following attempts to get rid of the fungus would be least effective?

    (1) powder the area between the toes
    (2) wear open sandals rather than closed shoes
    (3) wear socks that absorb moisture away from the feet
    (4) spray the feet with a prescribed fungicide
    (5) wrap the feet in a thick, tight bandage

Items 48–50 are based on the following passage.

In an attempt to enhance production of field crops, the state of Maryland established a free soil testing program through the state university. The state's agricultural extension service then advertised the free program to farmers. Specialists recommended types and amounts of fertilizers needed for maximum production of specific crops based on requested soil analyses. Utilizing the recommendations, farmers increased production, income rose, and the state received increased revenues.

Many farms required yearly applications of chemicals as the rainfall washed the dissolved nutrients into the rivers feeding the Chesapeake Bay. Farmers tilled more acreage, loosening additional soil particles that washed into the bay. Plant life in the bay flourished with the increased soil and nutrients, turning large areas into marshland. Algae clogged the waterways, and the excess nutrients upset the chemical balances in the water. This endangered the fish, oyster, and crab populations, and increased the price for bay fish and shellfish. Industries that previously thrived on bay fishing then failed and state revenues declined.

48. Which of the following statements best describes the message of the passage above?

    (1) Using science to solve practical problems is usually ineffective.
    (2) Sometimes the scientific solution for one problem presents another problem.
    (3) The world's food problems will never be solved.
    (4) Farmers should not use fertilizers to enhance food production.
    (5) Farmers should pay for cleaning the Chesapeake Bay since they caused the problems.

49. Which of the following results was independent of the use of chemicals?

    (1) increased state revenues from farmers
    (2) decreased state revenues from the seafood industry
    (3) increased plant life in the bay
    (4) decreased crab and oyster populations in the bay
    (5) increased soil testing by the farmers

50. Which of the following was least hurt economically from the use of chemicals?

(1) the state
(2) the bay fishing industry
(3) the farmers
(4) the bay seafood restaurant businesses
(5) the domestic shellfish consumer

Items 51–52 are based on the following passage.

When he was a student, Charles Martin Hall became interested in the extraction of aluminum metal from the ore bauxite when his chemistry teacher remarked that one might make a fortune from the discovery. From ancient times down through the Bronze and Iron Ages, metals were obtained by heating rocks called ores. For aluminum and many modern metals, the temperatures required to obtain the metals are too high for cost-effective extraction. After three years of investigating the problem, Charles Hall discovered that electricity could be used to release aluminum from its ore. Today, many modern metals are extracted from their ores by electrolysis.

51. Which of the following developments was most responsible for the availability and use of modern metals?

(1) heat-resistant furnace liners
(2) heat from nuclear reactors
(3) discovery of magnetism during the Iron Age
(4) discovery and production of electricity
(5) international space exploration

52. Ores are

(1) a class of metals
(2) inexpensive metals
(3) rocks that contain a metal
(4) metals brought to refineries by boat
(5) the wastes from blast furnaces

Items 53–54 refer to the following diagram.

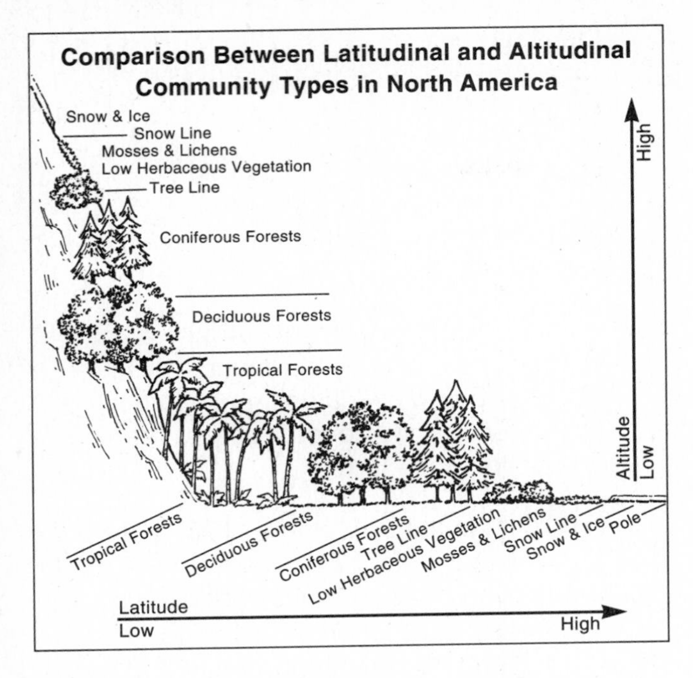

53. Which of the following conclusions is not supported by the diagram?

    (1) Altitude and latitude determine basic plant life type.
    (2) Deciduous forests are found in altitudes higher than tropical but lower than coniferous.
    (3) Plant life is limited to small forms at high altitudes and latitudes.
    (4) Tropical forests occur when both altitude and latitude are low.
    (5) Mosses are found in all forests despite latitude and altitude.

54. All of the following situations would result in a deciduous forest except when

    (1) both latitude and altitude are high
    (2) both latitude and altitude are moderate
    (3) the altitude is low but the latitude is moderate
    (4) the latitude is low but the altitude is moderate
    (5) either the latitude or the altitude is moderate

55. The maintenance and balance of environmental conditions is critical for the survival of living organisms. Which of the following examples least supports the above statement?

    (1) Many dinosaurs froze at the onset of an ice age.
    (2) Humans become ill when body temperature varies significantly from 98.6°F.
    (3) Crabs and oysters cannot reproduce in either fresh or salt water, but only in the mixture of fresh and salt waters in estuaries.
    (4) Boiling kills most microorganisms on surgical implements.
    (5) By mutation and variation of genetic material, many mosquitoes are no longer affected by DDT and other insecticides.

56. Scientists have been able to germinate seeds taken from the Egyptian pyramid tombs. Which of the following is an appropriate conclusion to draw from the above statement?

    (1) Plants live longer than animals.
    (2) Animals do not have food storage as part of the reproductive process.
    (3) Plant reproduction is superior to animal reproduction.
    (4) Animals that hibernate could also be preserved for long periods.
    (5) Some seeds are able to survive during long periods of dormancy.

Item 57 refers to the following information.

There are two types of butane. Both have the formula C_4H_{10}. Normal butane boils at 0°C, while isobutane boils at –12°C. Chemicals such as these with the same molecular formulas but with different properties are called isomers.

```
   H  H  H  H
   |  |  |  |
H— C— C— C— C —H
   |  |  |  |
   H  H  H  H
```

Normal Butane

```
        H
        |
      H-C-H
   H    |    H
   |    |    |
H— C —  C —  C —H
   |    |    |
   H    H    H
```

Isobutane

57. The difference in the boiling points of normal butane and isobutane is likely due to the different

(1) kinds of elements
(2) formulas
(3) number of hydrogen atoms
(4) number of carbon atoms
(5) arrangement of the atoms

58. Some musical instruments require a vibrating part to produce sound. Which example below indicates that a vibrating instrument part is <u>not</u> essential to the production of sound?

(1) rubbing the bow over a violin string
(2) hitting the skin of a drum with a stick
(3) crashing two cymbals against each other
(4) striking a piano string with a hammer by depressing a key
(5) blowing air through a bass tuba

Items 59–61 refer to the following diagram.

The Rack-and-Pinion Steering System

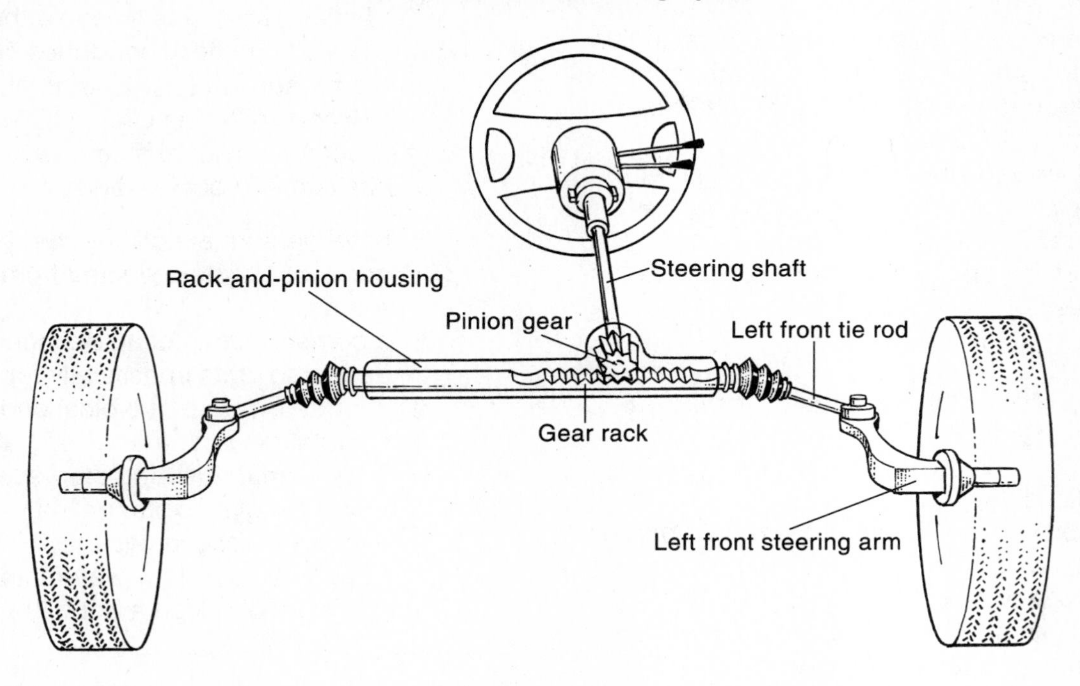

59. Rack-and-pinion steering operates by a series of

    (1) circuits
    (2) pulleys
    (3) nuts and bolts
    (4) gears
    (5) electronics

60. The purpose of the steering shaft is to

    (1) transfer steering movement to the pinion gear
    (2) move the steering arms up and down
    (3) hold the rack-and-pinion housing
    (4) join the left and right rods
    (5) connect the horn to the steering wheel

61. The person most likely to need an understanding of a rack-and-pinion mechanism is a

    (1) research scientist
    (2) long distance truck driver
    (3) race car driver
    (4) car mechanic
    (5) gas station attendant

Items 62–65 refer to the following information.

Receptor cells in the nose respond to gases; while those on the tongue respond only to liquids. The concentration of the chemical in the gas or liquid as well as the time and amount of exposure around the receptor cells determines the intensity of the taste or smell.

Saliva dissolves small amounts of solid foods which then in solution are exposed for taste identification. Solids and liquids can be smelled when molecules on the surface evaporate to become gases.

62. To experience the most taste from a liquid, a person can

(1) dilute the liquid
(2) collect extra saliva in the mouth prior to drinking
(3) swallow large quantities quickly
(4) swish or roll the liquid around the tongue prior to swallowing
(5) hold the nose closed while drinking

63. The presence of which of the following substances could least be detected by receptors in the nose?

(1) the burning of beans on the stove
(2) the fruitiness of a very ripe banana
(3) the barbecue being grilled at a backyard party
(4) the newly cut grass while mowing
(5) the hard candy in a covered glass dish

64. Microorganisms often live on other organisms, producing gaseous odors as they process products on or from the living or previously living host. Which of the following odors is least likely to be the result of microorganism metabolism?

(1) underarm perspiration
(2) aftershave lotion
(3) rotting onions
(4) dead mouse
(5) previously wet baby diaper

65. The composition and odor of perspiration varies from person to person depending on the type of foods eaten, chemicals applied to and microorganisms living on the skin, sexual and regulatory hormones secreted, and the presence of disease or infection. Which of the following circumstances does not illustrate individual differences in composition and odor of perspiration?

(1) hotel personnel noticing the different odors of sheets collected from different rooms
(2) a company marketing deodorants and antiperspirants specifically for men
(3) a nurse aware of a typical odor on a hospital gown of a feverish patient
(4) a vegetarian disliking the necessity of sharing a gym locker with a meat eater
(5) a teenager exceeding the recommended daily water intake after playing football on a hot day

Item 66 refers to the following illustration

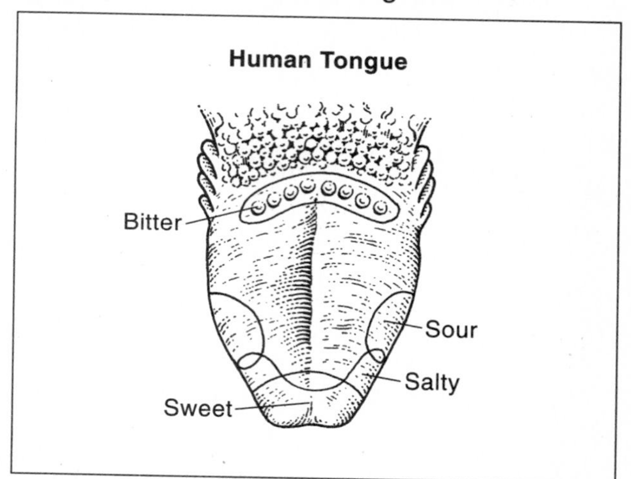

66. A mother has to give a bitter liquid medicine to her child. To minimize the bitter taste sensation, the mother might

(1) place the spoon to the right side of the mouth
(2) place the spoon to the left side of the mouth
(3) place the spoon over the back portion of the tongue
(4) place the spoon just over the tip of the tongue
(5) not use a spoon but let the child sip the medicine through a straw

Answers begin on page 89.

Analysis of Performance: Science Simulated Test B

Name: ______________________ **Class:** ______________ **Date:** ______

The chart below will help you determine your strengths and weaknesses in reading comprehension and in the science content areas of biology, Earth science, chemistry, and physics.

Directions

Circle the number of each item that you answered correctly on the Simulated GED Test A. Count the number of items you answered correctly in each column. Write the amount in the total correct space of each column. (For example, if you answered 24 biology items correctly, place the number 24 in the blank before out of 33.) Complete this process for the remaining columns.

Count the number of items you answered correctly in each row. Write that amount in the total correct space of each row. (For example, in the comprehension row, write the number correct in the blank before out of 15.) Complete this process for the remaining rows.

Test B Analysis of Performance Chart

Item Types:	Biology (Unit 1)	Earth Science (Unit 2)	Chemistry (Unit 3)	Physics (Unit 4)	Total Correct
Comprehension	1, **3,** 10, **14,** 15, 27, 46, 50	31, **34, 35, 39**	52	**21, 59**	____ out of 15
Analysis	**4,** 6, 11, **43,** 49, **53, 54,** 56, 62	32, **38, 40**	17, 19, **57**	**22, 23, 24, 60**	____ out of 19
Application	2, 9, **13, 28, 29,44,** 45, 63, 64, 65, **66**	33, **36, 37**	5, 7, 18 20	**25,** 26, 58	____ out of 21
Evaluation	12, 16, 47, 48, 55	**30**	**41, 42,** 51	8, **61**	____ out of 11
Total Correct	____ out of 33	____ out of 11	____ out of 11	____ out of 11	Total correct: ____ out of 66 1–50 = You need more review 51–66 = Congratulations! You are ready!

(**Boldface** items indicate those questions with a graphic stimulus.)

If you answered less than 51 questions correctly, determine which areas are hardest for you. Go back to the *Steck-Vaughn GED Science* book and review the content in those specific areas.

In the parentheses under the heading, the units tell you where you can find specific instruction about that area of science in the *Steck-Vaughn GED Science* book. Also refer to the chart on page 3.

Answers and Explanations

UNIT 1: BIOLOGY

Pages 4–14

1. **(2) bear, dog, beaver** (Comprehension) The tracks of the bear, dog, and beaver are the only ones to show marks beyond the ends of the toes that would be made by toenails.

2. **(2) bear** (Analysis) Bear tracks most resemble human tracks except for the claw prints. Beavers have webbing, hares and dogs have only four toes, and the muskrat toe proportion to foot length is unlike a human's foot.

3. **(3) webbing between the toes** (Application) Webbing between toes is an animal adaptation for swimming, which is necessary for building houses in midstream. The absence of pads or toes and the number and length of toes would not increase the ability to swim.

4. **(4) soft soil** (Application) Long, pointed fingers are good for digging soil. The absence of a grabbing thumb and absence of gripping claws make tree and chimney climbing improbable. No webbing which would be needed for living in midstream is indicated.

5. **(3) Beavers are larger than muskrats.** (Analysis) The beaver track is larger than the muskrat track, so if the tracks are drawn to scale it is likely that beavers are larger than muskrats. There is no evidence in the illustration to support any of the other statements.

6. **(4) Many ancient people used leather garments; therefore, the processing is simple and easy.** (Comprehension) Processing leather from hides was not easy in ancient times nor is it today. Without the multi-step process using chemicals such as salt and plant extracts, the bacteria would soon disintegrate the skins.

7. **(1) pelts to fur coats** (Application) Pelts (furry animal skins) are processed into fur coats (clothing) in much the same way as hides are processed for leather, except the hairs are not removed. None of the other options indicate an animal skin that becomes clothing.

8. **(4) the use of shoes as clothing** (Analysis) People would continue to wear shoes with no increase. The cost of leather products would increase as would the use of less expensive materials such as cloth and synthetics.

9. **(1) the skins for leather are a by-product from animals that have already been killed for their meat** (Evaluation) Most people accept the killing of animals for meat and do not have strong feelings about the hides being used for leather. The use of furs is not recent. Animals do not suffer more when killed for fur. Protests are against unnecessary killing, not against killing based on beauty. The manufacture of footwear does not require leather. Other products such as wood, cloth, and synthetics can be used.

10. **(3) The allergy is due to the chemicals used in processing leather.** (Analysis) The allergy is to the chemicals used in processing. The death of the animal and its skin or the removal of a part (hairs) would not cause an allergy. Tanning prevents bacteria from living on the leather.

11. **(5) data regarding the production of oranges in metric tons** (Analysis) Data from the graph considers only apples and peaches, not oranges. All other options can be verified by the graph.

12. **(1) humans** (Comprehension) Humans only assist pollination when natural agents are missing. All other options are considered natural agents.

13. **(5) wind** (Application) Wind needs no scent, nectar, or petals to attract it to be used as a pollination agent. Wild plants in nature do not rely on humans.

14. **(1) Male and female flowers grow on separate plants.** (Application) Option 5 is ruled out because all fruit-producing plants reproduce by flowers. The question implies that two or more plants may make fruit, but one definitely won't. This rules out option 4 because human assistance is not what makes the difference between producing and not producing fruit for this plant. Among the remaining options, only 3 describes the only situation in which one plant won't make fruit but two might.

15. **(5) produce the reproductive elements necessary for seed formation** (Comprehension) Flowers are the reproductive organs of complex plants that produce seeds. Increasing nutrient levels or beauty are not the functions of flowers.

16. **(5) the parent plant is healthy** (Analysis) A healthy plant is the factor related to healthy seeds. The amount of ovules, size of fruit, and the amount of complete flowers are individual plant characteristics. Pollen reaching the female structure does not insure good health for the seed.

17. (4) in a commercial greenhouse (Analysis) Plants in greenhouses do not have access to natural agents of pollination. Nonflowering plants do not need pollination. Farms, deserts, and backyards are accessible to natural agents.

18. (2) large, white, and fragrant (Application) A moth is most likely to find a flower that it can both see and smell. Therefore, moth-pollinated flowers are most likely to be white, large, and fragrant. Bright colors aren't useful at night because colors can't be seen in dim light.

19. (2) high in water content and low in fat (Analysis) Fats have high caloric values. Water has no caloric value. Potatoes are high in water and low in fat content. Potatoes have only 2.2% protein. They are not low in water. Whether a food is a vegetable or fruit does not indicate caloric value. A food can be low in calories and still have food value or be low in food value and contain many calories.

20. (4) cutting a potato into sections each having at least one eye, and then planting the sections (Comprehension) New potato plants come from sprouted buds that are located at the eyes. No other option includes the eyes. Potato plants do produce seeds but they are usually only used to crossbreed to obtain new or improved varieties.

21. (4) carbohydrate content increases (Application) Since oil is a fat, this cooking method will add fat. Since the oil is very hot, it will also boil away some of the water in the potato. Cooking in oil will not change the amount of protein, carbohydrate, or ash.

22. (3) 21.7 grams (Application) Water is 78.3% of a potato. 78.3% of 100 grams is 78.3 grams. Therefore, removing all the water from a 100 g. potato means removing 78.3 grams of its weight, leaving 21.7 grams.

23. (1) stingray (Analysis) Fish D is a stingray, because it is the only fish that is saucer shaped.

24. (4) weever fish (Analysis) Fish B is a weever fish, because it is not saucer shaped. It does not have narrow vertical body stripes. The mouth is not near the bottom of the head. There is a lateral horizontal stripe parallel to the dorsal fin.

25. (3) C (Analysis) Fish C's mouth is near the bottom. It has no narrow vertical body stripes and is not saucer shaped.

26. (5) Queen Anne's lace beautifying the roadside of an interstate highway (Application) The Queen Anne's lace is growing in an acceptable place. In all other options the plants were growing in places that resulted in situations that people or animal would not want.

27. (4) Many of today's weeds were the prized plants of colonists. (Comprehension) Many chemicals used by the colonists were grown in herb gardens and are now considered weeds. The idea of once a weed, always a weed is not correct. If a plant has a use, it is grown intentionally and not considered a weed. That many weeds were brought from foreign countries is a fact but is not the main idea of the passage.

28. (5) heavy spraying with unregulated chemicals that poison weeds (Evaluation) Heavy chemical spraying can be unsafe depending on the chemical used. The methods in the other options do not pose a possibility of harm to a person eating the food from that crop.

29. (3) dependent on the products and services of camels (Comprehension) The table indicates that some desert people depend on camels for work, food, clothing, shelter, energy, and transport needs. The table does not indicate any threat from camel overpopulation or competition. Eliminating camels would not increase survival which is dependent on, not independent of, camels.

30. (4) food supply is abundant (Analysis) Since the hump is fat, food must be abundant to make the fat. The hump is not water. The age, pregnancy, or sex of the camel does not affect the amount of fat.

31. (4) ability to survive in dry climates (Analysis) Since camels sweat very little, water does not leave the body, thus aiding survival in dry climates. Water and salt needs are decreased as they remain in the body. The ability to spit or reproduce is not related to sweating.

32. (2) prevent sand from entering during sandstorms (Analysis) Eyelids serve to protect eyes from objects such as sand entering. Animal eyes are not dry. The eyelids do not affect sleep or sight. Appearing to be blind would not help protect a camel.

33. (3) thick, tough skin (Analysis) Thick, tough skin would keep water from passing in or out. Many leaves would lose water to the air. Spines and thorns are protective devices, not water regulators. Thin skin would allow passage of water. Flower size is not related to water entering or leaving the plant.

34. (1) Tt and Tt (Comprehension) The individual genes shown on the top and left edges of a Punnett square make up the genotypes of each parent.

35. (2) 9 (Analysis) If T is dominant, then any offspring with a T gene will show the dominant trait produced by this gene. The Punnett square shows that the ratio of offspring with the T gene to offspring without it will be 3:1. If there are 12 offspring, the ratio 3:1 becomes 9:3.

36. **(4) half Tt and half tt** (Application) Half of the sex cells of the Tt parent will contain the T gene. Since this is the only source of the T gene, half the offspring will get it and be Tt, and half won't and will be tt.

37. **(3) two sickle-cell carriers.** (Application) For a child to get two recessive genes, one must come from the father and one from the mother. Therefore, both parents must have at least one copy of the gene. The only couple for which this is true is the couple in which both parents are carriers.

38. **(3) the sugar in fruits goes directly to cells without being changed to glucose** (Comprehension) The passage states that all sugars are changed to glucose before they are transported to the cells, so fructose does not go directly to the cells. All other options are correct as indicated by the passage.

39. **(3) mannose** (Comprehension) Mannose is a sugar. The names of all sugars end in the letters -ose.

40. **(1) assume the products have no sugar** (Evaluation) Consumers often assume the products have no sugar if the word sugar is not included in the ingredient list. Foods are not usually bought because they appear scientific or lack a common name. Names on labels do not cause illness.

41. **(4) the microorganisms that digest cellulose do not live in the human digestive system** (Comprehension) Microorganisms that live in the stomachs of ruminants are not able to survive in the human stomach; thus, humans cannot get energy from cellulose. All other options are untrue or are not supported.

42. **(5) a single pouch** (Analysis) The diagrams show the human and not the ruminant stomach to have only one pouch. All other options are ruminant differences.

43. **(4) Some bacteria are harmful to humans, but others are helpful.** (Comprehension) There are both harmful and helpful bacteria. That bacteria are more beneficial than harmful is an opinion not suggested by the passage. That both plants and animals can get bacterial diseases is a point made by the passage but does not summarize the passage. Bacteria are harmful to humans when bacteria are responsible for disease.

44. **(2) producing insulin needed by diabetics** (Comprehension) Producing insulin is a helpful function of some bacteria. Cabbage rot, salmonella, and anthrax are caused by bacteria, not killed, purified, or cured by bacteria. Bacteria do not poison the pea family but live in the nodules in a beneficial relationship.

45. **(5) accidents to death** (Analysis) Accidents are not a disease. All other options as in bacteria to pneumonia show a cause and then a disease. Cigarettes, viruses, parasites, and salmonella bacteria can all cause disease.

46. **(5) has eight legs and two body sections** (Comprehension) The spider does not meet the two qualifying characteristics of insects because it has eight instead of six legs and only two body segments instead of three. That legs face forward or have joints or that the creature has eyes on top or a big abdomen is not relevant to the qualifying characteristics.

47. **(4) is limited by inheritance but influenced by environmental factors** (Comprehension) The ability of an individual to excel at sports is limited by the genes inherited but depends also on environmental factors. Options 3 and 5 are incorrect because exercise and food are only environmental. Option 1 is wrong because athletic ability is limited by the genetic structure of the human body. Option 2 is incorrect because athletic ability is somewhat determined by the environment.

48. **(3) Is psyllium grown by regular farming methods?** (Evaluation) The method of farming is least important compared to whether the person is in need of fiber, whether it may have harmful side effects, whether it is as effective, or whether it is needed if other food adjustments can be made.

49. **(4) inability to know that a problem exists** (Evaluation) Elderly people certainly know there is a problem as they are the primary consumers of over-the-counter medications for irregularity. As people age, many bodily functions cease to operate as efficiently. Many elderly persons lack exercise, lack interest in cooking and eating full regular meals, or lack the money to purchase sufficient high-fiber foods.

50. **(3) misinformed** (Evaluation) Misinformed means that a person has not understood an idea correctly. The salesperson could be intelligent, overeducated, interesting, or could have scientific ability and still have a mistaken idea.

51. **(3) natural substances are not chemicals** (Comprehension) That natural substances are not chemicals is false. The salesperson is using this false idea to support the sales pitch. Some chemicals may cause cancer. Both natural chemicals and those changed by humans can be dangerous. The idea of chemicals can be used scientifically or otherwise. Reference to these facts, though, is not the basic wrong idea used to support the sales pitch.

52. **(5) metal alloys** (Application) An alloy is a mixture of two or more metals. Metal alloys do not contain life or oxygen. Ocean water contains oxygen which fish and other sea creatures use. Topsoil contains oxygen which bugs, worms, and microorganisms use. The atmosphere contains oxygen which land animals use. Hospitals increase the oxygen content of air for patients with respiratory problems.

53. **(2) whales and swordfish** (Comprehension) Whales and swordfish are not crawlers on the ocean floor as are all other options.

54. **(5) all ocean animals** (Application) All ocean animals depend on the phytoplankton for food either directly or indirectly. Therefore, all ocean animals would be affected if the phytoplankton died off.

55. **(4) keep them near the sunlight** (Analysis) Seaweeds are plants that need sunlight to make food. The sunlight is strongest at the surface. Seaweeds do not swim or need to breathe air. They make their own food and do not trap phytoplankton or animals for food.

56. **(3) is a deep water fish** (Application) Dark-skinned fish usually come from deep water. The skin color does not determine what a fish eats. River fish are usually light-skinned as rivers are generally not very deep. The question identifies the creature as a fish. How fish are prepared is not dependent on skin color.

57. **(2) fishers** (Evaluation) When fishers wish to catch a particular species of fish, they must know how deep to set their nets. All other options are not activities that require knowledge of the depth location of fish for success.

58. **(3) It is not very dark in deep ocean water.** (Analysis) Since very little sunlight reaches the bottom this statement is incorrect. The water on the bottom is not heated and is therefore much colder than the surface. Plankton use light to make food. To be considered a fish, a creature must have gills and be able to swim. Fish in deep water do not need eyes as there is not enough light to see in the dark, deep levels.

UNIT 2: EARTH SCIENCE

Pages 15–21

1. **(5) trilobite** (Comprehension) Trilobites are pictured in the lowest layer which was identified in the passage as being the oldest. All other options are in layers above the trilobite.

2. **(5) from hard parts of plants and animals** (Analysis) Teeth, shells, wood, and coral skeletons are the hard parts of plants and animals. Most of the fossils in the drawing are animal parts, but wood is from trees and plants. Trees and most dinosaurs with teeth lived on land while starfish, coral, and clams are sea animals.

3. **(2) deep on the ocean floor** (Analysis) Layer C contains fossils of coral which live in the ocean.

4. **(5) Clams were the first kind of animal to evolve in the oceans.** (Analysis) There are several fossils in the illustration older than the clam fossils, and several of these are of ocean-dwelling organisms; therefore, clams are not likely to have been the first ocean animals. The other statements are all consistent with the evidence in the illustration.

5. **(1) It was covered by the sea for a long time, then became dry land, then was covered by the sea again.** (Application) The oldest four layers of rock have ocean fossils, layer E has land fossils, and the most recent layer, layer F, has ocean fossils. Option 1 is the only statement that matches this evidence.

6. **(4) dry and cool** (Comprehension) Air from central Canada has been primarily over land, at a colder latitude, and is dry and cool. Wet and humid air develops over oceans, eliminating options 1, 2, and 3. Since Canada is farther north of the equator than the U.S., air from Canada is not usually hotter than air in the U.S.

7. **(3) tropical maritime** (Analysis) Tropical maritime air masses develop near the equator where the increased heat causes great amounts of water to evaporate. To form a hurricane, both heat and water are needed in the air. Polar regions do not contain sufficient heat, and continental regions cannot supply sufficient water for hurricane development. Hurricanes do not develop in cold or dry air masses.

8. **(2) in the uppermost layer of soil** (Comprehension) Earthworms will live where they have the most food, which is the top layer of organic material (decaying plant material is the major form of organic material in soil).

9. **(4) in a forest** (Analysis) Soil comes from the decay of plants and from the weathering of the parent rock. Therefore, all other things being equal, the deepest soil will be produced where plants grow well and where the parent rock can be easily weathered. This rules out options 1, 3, and 5. In addition, thick soil will not occur where it is likely to be eroded away, such as on a mountainside.

10. **(2) a soil with thick layers of organic material and silt and clay** (Application) The passage states that the upper layers contain most of a soil's nutrients, and the diagram shows plant roots growing through both of these layers. Therefore, a soil with thick layers of both organic material and clay and silt is likely to be the most fertile.

11. **(1) Indiana was once covered by an ocean** (Analysis) If limestone is found, consisting of dead sea creatures, an ocean must have covered that area. A tidal wave brings surface water on land, not ocean bottom rock or soil. Although rivers flow underground, the oceans do not extend underground to the middle of continents. Limestone does not come from desert areas. By definition, fish or sea creatures live in water with only brief stays on land.

12. **(4) the sand that formed the rock was blown there from different places** (Application) Layers in sedimentary rock are always the result of different sediments gathering in layers before the rock formed. Knowing that wind can carry sand from place to place, it is reasonable to assume that over time, winds carried and deposited different colors of sand to the place where the rock later formed.

13. **(5) plastic** (Application) All the other options are made from cement or glass, which require limestone rock.

14. **(3) sandblasting with air guns** (Analysis) Sandstone is cleaned by air guns that remove the outside particles. Because sandstone is so absorbent, liquids cannot be used to clean it. Paint, water, and varnish would seep into the rock. Wallpaper is not for exteriors.

15. **(3) hand lotion** (Application) Hand lotions contain oils and softeners to smooth skin. All other options contain hard or gritty particles.

16. **(3) a three-inch rise in the level of the oceans caused by partial melting of the polar ice caps** (Evaluation) Because global warming is global and long-term, the local or short-term evidence described in options 1, 2, 4, and 5 cannot be convincing. The polar ice caps could melt only if global warming were occurring.

17. **(4) the minerals and impurities they contain are different** (Application) The color of a gem depends on the mineral of which it is composed and the impurities in the mineral. The scarcity, value, hardness, or gem quality of a stone does not cause its color.

18. **(1) small** (Comprehension) Agates are not necessarily small, but to be semiprecious by definition, means a stone is beautiful, plentiful, colorful, and of less value than gems.

19. **(3) copper and iron come from rock materials** (Analysis) By definition, metals come from ores which are rocks. Since copper and iron are metals, they must originate in rock materials. All other options may or may not be true but cannot be derived from the information given.

20. **(3) dissolves in rainwater and is carried by rivers to the oceans** (Analysis) Salt is dissolved by rainwater and is carried to the oceans by rivers. Salt mines are inland, not on ocean bottoms. At ordinary air temperatures, salt does not evaporate. Salt is a mineral from rocks, not a product of animals. Salt is used by humans and animals to control body processes. It is not a waste of sufficient quantity to cause the salinity of the oceans.

21. **(3) elevation** (Application) Since air pressure changes with elevation, measurements of air pressure could be converted to measurements of elevation.

22. **(2) the pressure of the atmospheric air increases as the elevation decreases** (Analysis) The air pressure increases as elevation decreases. When the air column is greater, the weight of air pushing on the balloon is greater. Since the balloon and the air inside can expand or contract, the balloon becomes smaller because the air pushes in on each square inch of surface with greater pressure.

23. **(5) Trees are unable to grow on high mountain tops where air pressure is decreased.** (Application) The inability of trees to grow at high elevations is not dependent on air pressure, but on temperature and soil depth. All other options contain an item that can expand or flex in response to increased or decreased pressure.

24. **(4) The sun sends more rays to land than it does to water.** (Comprehension) The land absorbs more of the energy than does the water. The drawing indicates that air is rising over the sand but not over the water. Thus the sand is hotter, has a higher temperature, and causes an inland breeze. The cooler water causes the air to sink.

25. **(4) C and D** (Comprehension) California had greatest number of quakes, and Colorado had no high-intensity quakes. Hawaii, an island, does have many quakes. New York, on the east coast, had 52.

26. **(5) Washington** (Analysis) Washington has the highest (1:14). To find the proportion, the number of high-intensity quakes is divided into the total number of earthquakes. California has 1 in 28, Colorado has 0 in 59. North Dakota has 0 in 0, and Utah has 1 in 34.

27. **(1) in, near, or bordering the Pacific Ocean** (Analysis) The states that are in or border the Pacific Ocean have the most quakes: California, Alaska, Hawaii, and Washington. No southern, Gulf, Great Lakes, or Great Plains states are listed. The eastern states listed have many fewer quakes.

28. (2) wind (Comprehension) Winds and a source of sand cause desert dunes. In deserts the sand is already mostly on the ground, eliminating gravity. Glaciers are not in the tropics. Deserts do not have ocean waves or much running water to cause the huge dune piles inland.

29. (3) June 21 (Analysis) The longest daylight is in summer. June 21 is the only summer date listed. January and December have short days and March 21 is spring equinox, with equal day and night hours.

30. (2) burning fuels (Application) Cars, trucks, and planes all burn fuels to obtain the energy needed to move. Some researchers have attempted to outfit moving vehicles with solar cells. Nuclear, geothermal, and tidal energy still require a large facility, are restricted to a particular place, or require wires to transmit the energy as electricity.

31. (2) fossil fuels (Comprehension) Fossil fuels are made from the materials of living organisms that lived in the past. All other options have an accessible energy supply for many future generations to use.

32. (5) There would be no leap year day. (Comprehension) The thirteen month calendar does have a leap year day (June 29). All the other options are characteristics of the calendar.

33. (2) calculating daily profits quarterly (Evaluation) If profits were calculated quarterly, one month of 28 days would not be accounted for. Persons on monthly salaries would gain one extra month's pay. Federal holidays would always fall on the same date and day of the week and would be easier to remember. All countries would use the calendar for business, easing communications for international transactions, and military operations.

34. (4) length of a year (Analysis) The rotation of Earth determines the length of a day (24 hours) and the revolution of Earth around the sun determines the length of a year (a little over 365 days). All the options except option 4 list things that humans have decided on and that could be different.

35. (4) because 365 divided by 13 has a remainder of 1 (Analysis) If a year is divided into 13 equal months, one day is left over. This day is accounted for by "year day."

UNIT 3: CHEMISTRY

Pages 22–27

1. (1) only ebony (Comprehension) Ebony is the only wood with a density greater than water. Therefore ebony would sink in water, while all the others are less than 1 gm/cc and would float.

2. (3) road contractor (Evaluation) The small amount of wood used in road building does not depend on its density or buoyancy. Boatwrights must construct boats that are not only strong but will float. Woods and materials heavier than water can be used, but the denser the material, the smaller the load a boat can carry. Carpenters, cabinet workers, and apartment construction supervisors choose woods not only for strength and durability but for ease in construction.

3. (1) pound (Application) Weight is directly dependent on density. All other prices would not be affected by density; thus the price for all other options would be the same for any wood type.

4. (3) of 0.6 g/cm^3 or greater. (Analysis) Pine has a density of up to 0.6 g/cm^3, so the definitions in options 1 and 2 would include some pine wood as hardwood. The definition in option 4 excludes some hickory, oak, and maple from the hardwood category. Option 5 doesn't make any sense if there are only two categories of wood. The definition in option 3 is the best choice since 0.6 g/cm^3 is the dividing line between the densest pine and the least-dense hardwoods.

5. (4) Humans have walked on the moon and found no atmosphere there. (Analysis) The fact that the moon has no atmosphere does not support the presence of matter surrounding Earth's surface. The molecules in air help support airplanes. Wind and breathing give indirect evidence of the presence of matter. When a spacecraft reenters the atmosphere, the craft rubs against the molecules, causing heat from friction.

6. (5) makes water molecules at the surface more likely to break away from the rest (Analysis) Evaporation occurs when the molecules of a liquid break away from the surface of the liquid and enter the gas phase. Heating encourages this process because it gives the molecules of the liquid more kinetic energy, which makes them vibrate and move around faster.

7. (3) CCl_4 (Comprehension) the prefix tetra- refers to the number 4. The letter symbols for carbon and chlorine are C and Cl. Options 1 and 2 do not reflect the number 4. Co and Cu refer to cobalt and copper.

8. (1) carbon dioxide (Comprehension) The subscript 2 indicates two atoms of oxygen. The prefix for the number 2 is di-. There is no other number to indicate the other prefixes.

9. (1) CH_4 (Comprehension) CH_4 contains only carbon and hydrogen atoms, which is consistent with the definition of a hydrocarbon. All other options contain oxygen.

UNIT 3

10. **(3) The melting and freezing temperatures are the same.** (Comprehension) Melting and freezing occur at the same temperature, as shown in the diagram. They are simply phase changes in different directions.

11. **(1) stays the same throughout the time the water is boiling** (Analysis) The flat line at the boiling temperature in the phase diagram indicates that the temperature of the liquid water stays the same during boiling, even as more heat is added.

12. **(3) sorting** (Application) The series of sieves is a mechanical means of sorting fragments by size. All other options are wrong by their definitions.

13. **(5) gravitation** (Application) Fat is less dense than the flour/broth solution; thus it floats, separating the two substances. All other options are wrong by their definitions.

14. **(2) distillation** (Application) Distillation requires vaporization and condensation for separation. In this way, petroleum is separated to obtain the desired substances. All other options are wrong by their definitions.

15. **(1) extraction** (Application) Vanillin dissolves in a solvent, separating it from the bean. The use of a solvent to separate chemicals is extraction. All other options are wrong by their definition.

16. **(4) magnetic separation** (Application) Iron is magnetic and is separated by using magnets. All other options are wrong by their definition.

17. **(1) salt from sea water** (Application) Salt is a solid dissolved in a liquid. The other options don't list solids dissolved in liquids.

18. **(5) wool** (Comprehension) Wool ignites at 400°F. According to the graph, all other required temperatures are higher.

19. **(3) rayon** (Comprehension) According to the graph, silk's kindling temperature is about 1050°F, and rayon's is about half that, or 525°F.

20. **(2) a long-sleeved nylon shirt** (Evaluation) According to the diagram, nylon catches fire only at very high temperatures, and short sleeves would not protect the cook's arms. Long sleeves would protect the cook's arms. Silk is impractical.

21. **(2) leaving a six-pack of carbonated beverages in a car parked all day in the hot sun** (Application) Although the inside of the car will become hot and the cans may explode, carbonated beverages do not burn easily. When car engines, painting chemicals, and cloth overheat, fire may begin. Gasoline has a low kindling temperature which is why No Smoking signs are posted at gas stations.

22. **(1) the head is composed of two different chemicals** (Analysis) The fact that the head is composed of two chemicals is not a condition of ignition. In safety matches, the head has one chemical and another is mixed into the scratching surface. Kindling temperature for the chemical on the match head should be low, and the scratching materials should be rough. There must be sufficient oxygen. The match must burn with enough heat to ignite other substances.

23. **(4) the blanket prevents the burning item from reacting with the oxygen in the air** (Analysis) When wrapped around a burning object, a wool blanket prevents oxygen from continuing to react with the burning object. That wool has a low kindling temperature or is rough to the touch does not explain how it can be used to put out fire. Water to put out fire must be on the fire and not absorbed by the blanket.

24. **(2) comparing their natural colors** (Comprehension) If all the noble gases are colorless, comparing their color could not distinguish one from another. According to the table, each member has its own weight, density, and boiling and melting points that can be used for identification.

25. **(4) by liquefying air, then separating the liquid mixture** (Analysis) All gases except radon are found in the atmosphere, so the place to obtain them is from the air. All other options do not include air or the atmosphere.

26. **(1) escaped into space** (Comprehension) According to the passage, hydrogen is light; light atoms escape into space. It also says that large amounts of hydrogen were released into the original atmosphere, and that at present, very little hydrogen is in the atmosphere. Therefore, it is strongly implied that the original hydrogen escaped into space.

27. **(2) Hydrogen atoms float in the air and escape into outer space.** (Comprehension) Hydrogen atoms escape into outer space. The fact that water, fuels, fats, and carbohydrates contain hydrogen does not explain hydrogen's absence from air. Hydrogen's presence on the sun and stars does not explain its low concentration in Earth's atmosphere.

28. **(1) one of the atmosphere's heavier elements** (Analysis) Nitrogen sinks in air to Earth's surface because it is heavier than the gases which float. Nitrogen's reactiveness, absence from soil, and its use in life forms do not explain why it sinks.

29. **(2) dry-cleaning machine** (Application) Dry cleaning machines use various chemical solvents,

such as carbon tetrachloride, to dissolve dirt on fabrics that cannot be cleaned in water. Almost all liquids for human consumption contain water. Washing machines use water to dissolve dirt. Fertilizers get into plants by dissolving in water and being absorbed by roots along with the water.

30. **(2) a bag of mixed jellybeans** (Comprehension) Despite the color, there is only one substance—jelly beans. Alcoholic drinks, coffee, and fertilizer all contain dissolved materials in water.

31. **(5) Rats fed large amounts of the compound showed no ill effects over several generations.** (Evaluation) Option 5 describes the kind of evidence that may help convince scientists of a compound's safety. The evidence in option 4 is less convincing because the compound could have long-term effects that don't show up immediately. Options 1, 2, and 3 are irrelevant because even if they are true, the compound could still be harmful.

32. **(4) water freezes** (Application) Salt, a tree, and iron are all solids in which the atoms or molecules are arranged in a particular, orderly way. When the salt dissolves, the tree decays, or the iron rusts, this order is destroyed. In contrast, when water freezes, water molecules are arranged into a more orderly pattern.

33. **(5) will not change** (Comprehension) If a number is a constant, it does not change. All other options indicate possible change.

34. **(2) The concentration depends on the amount of acid.** (Analysis) Concentration is a dependent variable that changes as the amount of acid varies. Because the concentration was lessened, it cannot be a constant. The amount of acid is of concern to the question, not the amount of solution. A small amount of concentrated acid would not necessarily make a concentrated solution. The concentration was decreased, indicating a relationship to the acid.

35. **(4) is dependent on the surface area exposed** (Analysis) The amount of water left is dependent on the amount of water evaporated, which depends on the amount of surface area exposed.

UNIT 4: PHYSICS

Pages 28–34

1. **(4) 10 cm^3** (Comprehension) The graph shows that there is a 10 percent change in volume from 25° to 55°. Ten percent of 100 cm^3 is 10 cm^3.

2. **(1) expansion by heat is dependent on the kind of liquid but independent of the kind of gas.** (Analysis) The rate of the expansion of liquids depends on the type of liquid. Air is a mixture, but the other gases are not mixtures. The three gases were tested separately but yielded the same results as all three responded in accordance to Charles' Law. Water at 55°C increased to 4 cm^3; whereas petroleum at 55°C changed only to 3 cm^3. Both gases and liquids expand as temperature increases. Liquid expansion is individual, but gas expansion is the same for all gases.

3. **(4) The gases would expand in conformance to Charles's Law.** (Application) Gases from anywhere in the universe are expected to conform to the laws of nature. The origin of the gas and the conditions where it is found should not affect its conformance to universal laws.

4. **(4) increase by an unknown amount** (Analysis) All the lines on the graph show increases, so it can be inferred that the volume of a liquid increases when heated. It is impossible to know how much the increase will be without knowing exactly what the liquid is.

5. **(5) the air pressure in the thin glass tubes is falling** (Comprehension) The Bunsen burner flame indicates heating; the thermometer indicates the measurement of temperature; the shading indicates liquid matter; the changes in the fluid levels in the pipettes indicate expansion. There is no evidence to indicate a change in air pressure.

6. **(5) buzz saws for cutting timber** (Application) A buzz saw's movement to cut wood is not dependent on moving air but is dependent on burning a fuel.

7. **(3) sewing machines** (Application) A sewing machine's parts are moved by a motor that uses electrical energy. In all other options a fuel is burned.

8. **(1) rocket ships** (Application) Rocket ships have no pistons, wheels, or turbines that assist in movement as the other options have. The entire body, except for the area that releases the gases, is pushed from the inside.

9. **(2) wheelbarrow** (Application) In a wheelbarrow, a force is applied on the handles, the force is transmitted to the load, and the whole thing pivots on the wheel, thus meeting the definition of a lever.

10. **(4) when it is heated to extreme temperatures** (Comprehension) The passage says matter on the sun exists as plasma, but not that the sun is the only place plasma can exist. It says that solids, liquids, and gasses can all enter the plasma state. It says nothing about radiation or pressure.

11. **(5) in glowing stars** (Comprehension) All glowing stars, of which the sun is one, are undergoing nuclear fusion that produces intense quantities of heat energy that ionizes matter into the plasma state. No other options involve heat in the tens of thousands of degrees Celsius.

12. **(1) air in a balloon** (Comprehension) Air in the balloon has a shape, not of its own, but that of the balloon. The air remains a gas. All other options maintain their own size and shape as long as the temperature of the object does not change.

13. **(1) the energy stored in the steam was transferred to the movable turbine wheels creating motion, and the decrease in energy changed the steam to water** (Comprehension) Heat energy (stored in the motion of gas molecules) is transferred into the motion of the blades. Having lost heat, the gas returns to the lower energy state of a liquid. The steam acts on the turbine, causing it to respond by turning. The loss of energy occurs whether or not the turbine becomes warm. The blades, in moving contact with the steam, would increase the temperature. Evaporating liquids would decrease temperature.

14. **(3) Photovoltaic cells have some advantages over fossil fuel-burning power plants.** (Evaluation) All the conclusions except the one in option 3 require making assumptions or value judgments that don't relate directly to the information in the paragraph.

15. **(5) "I wonder how many ways the future generations will use these interesting rays?"** (Evaluation) The hope and optimism of finding uses for new information drive scientists to pursue further knowledge. All other options are comments that would have stopped or limited the many uses we now make of lasers.

16. **(1) cutting metal in industry** (Application) Cutting metal means destroying or changing matter. The other options list uses in which a laser is aimed at an object but the object is not changed or destroyed. It is logical to assume that destroying or changing matter requires the most energy and therefore the most powerful laser.

17. **(4) residual radioactivity** (Analysis) Survivors would need to prevent exposure to the falling particles. All other options are in the past in the first hour after an explosion. Survivors need only be concerned about what is yet to come.

18. **(2) flash** (Comprehension) Flash is an extremely bright light given off at the release of energy and causes blindness. All other options vaporize, break apart, kill, or in some way affect other body processes.

19. **(1) brass** (Application) Tubas and trumpets are made of metal, have valves to change the length of a vibrating air column, and are brass instruments. They do not have tight skins, reeds, strings, or holes to cover. All instruments must cause vibrations to produce sound. Vibrator is not an instrument group.

20. **(4) strings** (Application) Guitars and banjos have strings that vibrate. Folk is not an instrument group. Guitars and banjos do not have valves, tight skins, or holes to cover.

21. **(3) the gong** (Comprehension) A gong is a percussion instrument made of metal and must itself vibrate to produce sound.

22. **(2) B** (Application) Brakes give the force needed to stop the motion of the car. All other options are not an attempt to slow or stop an object in motion.

23. **(1) A** (Application) The heavy truck requires more force to change its motion, using more gasoline than the car with lesser mass.

24. **(2) Earth continues to revolve around the sun.** (Analysis) Example B illustrates part of the First Law, that an object will stay in motion unless a force acts on it. Option 2 is the only example illustrating the continuing motion of an object or the force required to stop its motion.

25. **(3) C and D** (Application) Both the boat and the spinner react in the opposite direction in response to an action. All other options are not responding by movement in the opposite direction.

26. **(3) a flashlight shining after exchanging the old batteries for fresh ones** (Application) The electricity in a flashlight is continuous, not one incident of discharge. All other options involve rubbing: clothes in the dryer, shoes on carpet, nylon against nylon, and clouds against clouds.

27. **(1) gold and silver** (Application) Gold and silver are used because the expense and means of repair are too costly and time-consuming to use with expensive space equipment. All other options are metals that are more likely to burn or need repair.

28. **(3) the flow of electrons is not continuous with static electricity** (Comprehension) Use of electricity depends on a continuous supply that static electricity cannot maintain. Static discharge or continuous flow is not dangerous when protected by fuses. The rubbing just builds up a charge and does not supply a constant flow.

29. **(4) the labor and repair contractors who repair the damage or demolish the structure and rebuild a new home** (Evaluation) The labor and repair contractors make money on fires started by improper fuse installation or usage. All other options result in harm or additional expense to those involved.

30. **(2) induction coil** (Comprehension) A van or bus has an induction coil as do cars and trucks to turn low voltage direct current to high voltage. Batteries supply only low voltage current. Transformers are designed for alternating current.

31. **(5) step down transformer** (Application) U.S. appliances are made for 120 voltage; when using appliances in countries with 240 voltage, a step-down transformer must be connected between the outlet and the appliance. Batteries and dry cells do not change voltage. Induction coils are used only with batteries. A step-up transformer would have increased the voltage even more.

32. **(5) lenses** (Analysis) Lenses focus light in glasses to help people see objects more clearly. Mirrors would keep light out. Prisms would produce rainbows of light which would interfere with vision.

GED SIMULATED SCIENCE TEST A

Pages 35–54

1. **(1) historical skeletal remains** (Comprehension) Skeletal remains are only bones. No fingerprints would be on a skeleton. Kidnapping and amnesia victims, workers, and dead bodies in a morgue would have fingerprints for possible identification.

2. **(4) to assist in grasping and holding objects** (Comprehension) Grasping and holding objects by hands is possible because of the ridges on the fingers. Otherwise, objects would easily slide away. The passage does not indicate beauty, cleanliness, or fluid retention. Identification is the use of the uniqueness of the ridge but is not the reason for having prints.

3. **(5) body measurements** (Application) Body measurements include height and weight, which are not totally observable in a facial picture. Genetic information, fingerprinting, and branding do not figure in measuring the height and weight of an individual.

4. **(3) genetic configurations** (Application) Genetic configurations could only be determined by a microscope. Sperm would not have fingerprints, voices, scars, or body parts to measure.

5. **(5) seals and walruses swim in cold ocean water** (Analysis) The ability of seals and walruses to swim in cold water does not give evidence that active life processes require heat. Human body temperature indicates heating. That life activity slows in seeds, lizards, and bears in colder times also indicates that heat is needed for normal activity.

6. **(4) the reactants and products reached equilibrium** (Analysis) If equilibrium was reached the symbol $\rightleftharpoons$ would have been used. The Δ symbol indicated that heat was needed. The + shows the separation of Hg from O. The ↑ symbol indicated that O was produced as a gas. The number 2 before HgO indicated that 2 mercury oxide molecules were needed.

7. **(1) a secretary using shorthand** (Analysis) The symbols used by chemists represent words and are used to quickly express ideas that can be written out in words, as a secretary would use shorthand symbols for writing words. Ingredients, goggles, paints, and gloves do not represent words.

8. **(1) A gas is produced.** (Comprehension) The symbol ↑ on the right side of (H↑) indicates a gas was produced. The equation does not use a symbol for metals, acids, or dissolving liquids. It does not show the symbol for heat (Δ).

9. **(5) a daily newspaper article about the discovery of a new compound to treat AIDS** (Evaluation) The daily newspaper does not usually show scientific equations because many people reading the paper are not familiar with the meanings of the symbols. Pharmacists, chemical engineers, research chemists, and chemistry students learn and use the chemical symbols.

10. **(3) 14** (Comprehension) Use the right side of the graph (grams per 100 m^3), then follow the bottom temperature until you find 10°C, look up the 10°C until it is crossed by the curve just beneath the 15 line, halfway between 10 and 20 on the right side. Thus, 14 grams is correct. The answer 11 grams is incorrect as it reflects ounces per 1000 ft^3, left side; whereas the question asked for $g/100m^3$. The graph does not indicate the other answers.

11. **(3) The information is presented in both English and metric units.** (Analysis) Although the English system is used in the United States and a few other countries, most of the world uses metric units. The graph provides information in both units for worldwide use. Whether a graph is scientific, is accurate, or lists the author does not make its use international. Water for different localities does not affect air's maximum holding capacity and does not increase the graph's international use if the units are not understood.

12. **(5) a deep-sea diver investigating sunken treasure** (Evaluation) the deep-sea diver works underwater and thus would not be affected by the amount of water in the air. Hair, food, and breathing are all affected by humidity. The weather forecaster must know the relative humidity in order to report it.

13. **(2) Many hybrids and pure breeds can no longer survive in the wild.** (Evaluation) When humans are unable or choose not to maintain the artificial environment, the hybrids and pure breeds may not survive. It is not harmful that the dog is a friend to humans, that 121 breeds exist, or that desirable traits are fixed. Plants and animals rely on many agents and conditions in evolving; the addition of humans is just one more agent and may have prevented extinction.

14. **(5) The dog is one of nature's survivors.** (Analysis) Being nature's survivor does not indicate artificial selection. Left to nature alone, modern dogs would not exist. Domestic implies that humans must maintain its environment. The great diversity of specific breeds indicates human's intervention. It takes a human to establish a system of pedigree. That humans are a primary agent supports the fact that artificial selection was utilized in dog evolution.

15. **(3) Humans must maintain the artificial environments necessary for the survival of the selected breeds and varieties.** (Evaluation) Since many domesticated plants and animals would die without the artificial environments, humans must maintain them or face extinction of the variety or breed. All other choices benefit humans without additional responsibilities.

16. **(3) inbreeding** (Comprehension) Hybridization has already been done. Pedigree is contingent on registering known ancestors of a breed. (A purebreed is established after inbreeding.) Artificial selection is more than just crossbreeding hybrid offspring; it includes all the other choices.

17. **(3) crossbreeding** (Application) In crossbreeding, the desirable traits are selected, then individuals with those traits are bred in order to produce offspring with the desirable traits. The farmer is using purebreds with the already inbred characteristics to crossbreed. A hybrid is not yet produced; it is hoped for and will be a result of successful crossbreeding. Hybridization is the result of cross breeding, while fertilization is any union of egg and sperm and occurs with or without crossbreeding.

18. **(2) selective intervention by humans** (Evaluation) Highly specialized parts for human consumption indicate human selection since these specific food storage parts are not basic to each plant's survival. The word domestic in the title indicates human intervention, not natural selection. Wind for seed dispersal does not develop specialized parts. Adaptations to climate would not cause the plants to become domestic. Ceremonial usage is unlikely to produce all the six different characteristics in food storage for humans.

19. **(4) root** (Comprehension) The roots of Brassica oleracea are not adapted for food storage or enhanced for consumption. Kale is a leaf, kohlrabi a stem, cabbage a bud, and cauliflower and broccoli are flowers.

20. **(2) Many edible wild species can propagate themselves.** (Analysis) That edible wild plants propagate does not give further evidence that Brassica oleracea cannot propagate itself. Dependence, artificial conditions, and co-evolution are results of domesticity. That nature allows only the fittest to survive also supports the idea that these plants survive only for and with the help of humans.

21. **(3) C** (Application) The no-effect graph would show that the power reactors have not changed conditions that endanger people. All other graphs indicated change.

22. **(1) A** (Application) The inversely proportional graph would show that as the spill covers more of the bay, less animal life remains. Directly proportional would indicate more animal life with more oil. Oil spills do affect animal life; the leveling-off, and peaking and tapering graphs both show an increase at the beginning of an action which does not occur with animals as the spill invades.

23. **(4) D** (Application) the temperature rose to a point and increased no further, indicating the leveling-off graph. As the temperature did not continue to change, inversely proportional, directly proportional, and peaking and tapering are eliminated. The no-effect graph is eliminated because heating did raise the temperature initially.

24. **(3) Certain people's bodies react to chemicals called allergens by producing histamines.** (Comprehension) The production of histamine in reaction to chemicals is an allergy. Shellfish is an example of an allergen, not the basic cause. Tightening muscles is a reaction, not a cause. Avoiding contact is a treatment. That allergens are found everywhere does not explain why only some people react.

25. **(4) Histamine produced by the body in response to allergens causes mucous glands to oversecrete.** (Analysis) Histamine causes mucus membranes to oversecrete, producing

SIMULATED TEST A

nasal drip. The production of mucus is caused by histamine-stimulating glands, not by the nose. Family history does not cause nasal drip. Glands, not capillaries, produce mucus. Avoidance of allergens does not explain why a nasal drip accompanies hay fever.

26. (1) avoiding foods with chocolate (Application) Avoiding an allergen prevents attacks. If the woman is still allergic, she still produces histamines. The allergen is found in chocolate regardless of its type. Antihistamines are used only for a severe attack. Worry or lack of worry does not prevent a chemical reaction.

27. (4) used to alleviate symptoms in severe attacks (Comprehension) Antihistamines are used when a severe attack occurs. Antihistamines are not the only treatment, but the allergen, not antihistamine, is used to desensitize persons. Antihistamine is a treatment, not an avoidance.

28. (1) The individual is only allergic to the pollen and not to other substances being tested. (Analysis) There is evidence only for the substances tested. There is no evidence that the person or his/her parents or family are allergic to dust, nuts, or any other type of pollen.

29. (1) Request for increased energy to move machine parts stimulates the body to produce muscle tissue from proteins. (Analysis) Muscle tissue is increased in response to the force demanded. Eating meat is not dependent on the machine. Decreasing fat or increasing oxygen or high energy foods does not increase muscle tissue. Increasing the proportion of muscle to fat does not increase actual muscle. Increasing appetite does not necessarily mean the person will eat protein, the food needed to build muscle.

30. (1) In most cases of malaria diagnosed in the United States, the individual contracted the disease outside the U.S. border. (Analysis) The U.S. is a light area on the map, where malaria has never existed or has been eradicated. Therefore, if a person in the U.S. contracted the disease, it was probably while traveling. Southeast Asia is dark on the map and has high incidence of malaria. The map does not indicate the direction of the spread of malaria, if any. Cold and temperate areas are not areas of high risk and are noted as light on the map. Much of Earth's densely populated areas are high-risk areas.

31. (5) In northern and southern Africa, the incidence of malaria is limited. (Evaluation) Since the traveler is going to central Africa, the incidence of malaria in northern or southern Africa is of no consequence. All other choices give the traveler information that can help prevent or recognize the onset of the disease and thus seek treatment quickly.

32. (2) a puncture wound while digging in the garden (Application) Soil and a deep wound are both present in the gardening puncture wound. All the other choices do not indicate both factors.

33. (4) a woman manually examining her breasts for possible tumors (Comprehension) No chemicals are involved when the breasts are inspected manually for possible tumors. Blood, urine, or stomach fluids are all chemicals that react with indicators. The breath contains gaseous chemicals that are analyzed by a breathalyzer.

34. (2) follow normal eating patterns (Comprehension) Tan over blue is normal and requires no deviation from the usual eating pattern. All other choices indicate a change, or action rather than no action, and the continuation of normal eating.

35. (3) biochemistry (Comprehension) Bio, meaning life, and chemistry, referring to chemicals, would refer to body cells, fluids, and gases. Engineering refers to chemicals in structures and machines. Astro refers to objects in space such as stars. Inorganic means nonliving. Organic refers to carbon chemicals only.

36. (3) light energy travels faster than sound energy (Analysis) An explosion produces light energy and sound energy at the same time; if the light energy reaches you first, then it must travel faster. Options 4 and 5 can be ruled out based on personal experience.

37. (2) placing protectors in unused electrical outlets (Evaluation) Protecting children from access to electrical shock increases their safety. Lowering air temperature and using humidifiers may be healthy but do not increase electrical safety. Gas appliances are not safer than electrical appliances. Leaving lights on in a child's room may have psychological value but does not increase safety.

38. (3) At the hydroelectric plant, a turbine shaft moves a wire coil through a magnetic field generating electricity. (Analysis) A power plant generates electricity. It does not change electricity to usable forms. Electricity is changed to heat needed to raise the temperature to ignite the gas stove, to movement of the fan blades, to sound in a radio, and to heat by a furnace.

39. (3) blenders and mixers (Application) Blenders and mixers involve motion of parts. All others require different forms of energy—light, sound, heat, or the removal of heat.

40. (5) an increase in the longevity of the appliance (Analysis) Long-term use is not increased by continuous use. Wearing on parts, increased bill, overheating, and increased need for maintenance can all result from continuous use.

41. (2) switch (Comprehension) Switches control when the electricity will be used. Fuses protect us from excess current which could cause fire. Transformers control voltage for transport and special usage. Insulators protect us against electric shock. Wires are used to transport current.

42. (5) a double cage to a lion in a zoo (Application) The insulation around a wire prevents a person from possible death by electrocution. The cage prevents death by the lion. A book cover still allows access to the book. Rugs, shades, and coats all surround an object but do not prevent access or death.

43. (3) conduction and convection (Application) Heat is flowing from the heating element to the pot by conduction; it is also flowing by conduction from the pot to the water in direct contact with the pot. Within the water, heat is flowing by convection.

44. (1) a mathematician calculating the weight of several objects (Evaluation) The mathematician calculating does not involve two objects that may hit each other. A boxer or football tackle is involved in hitting another person. The vehicle driver must attempt to prevent hitting other vehicles. The pilot directs the angle of take-off to get lift. Sufficient speed for the mass of the plane is important in liftoff.

45. (5) pressure on the back of the skull resulting in diminished vision (Analysis) The back of a skull contains no sinuses and is least affected. All other choices affect areas near the inflamed sinuses.

46. (5) sniffing a vapor that causes the mucous to flow (Evaluation) Vapors that increase mucus flow allow the sinuses to drain, reducing the pressure and pain. All other choices do not indicate that the trapped mucus would be released.

47. (3) family physician (Evaluation) The family physician screens patients and then treats them or refers them to an appropriate specialist. The doctors who generally treat sinus problems are ear, nose, and throat specialists and/or allergists. Surgeons perform operations, cardiologists are heart specialists, podiatrists specialize in foot problems, and dermatologists treat skin conditions.

48. (5) telescope (Application) In using binoculars, a person is attempting to make an object appear closer, which is what a telescope does. A microscope only enlarges small objects. The other choices do not cause objects to appear closer.

49. (2) oscilloscope (Application) A heart monitor is used to view the electrical impulses of the heart as visual lines on a screen, which is a type of oscilloscope. None of the other choices turn electricity into lines on a screen.

50. (1) microscope (Application) Bacteria are microscopic. In order to verify their presence, they are made to appear larger, but not closer, by using a microscope. None of the other choices magnify. Telescopes make objects appear closer.

51. (4) spectroscope (Application) A spectroscope identifies elements by analyzing colored light spectrums. The sun emits light. Other choices may use light but do not break light apart into a spectrum to identify elements.

52. (4) more of all light energy wavelengths (Analysis) A satellite is above or near the top of Earth's atmosphere. The illustration indicates more light of all wavelengths is received at the top of the atmosphere, not just more infrared. The sea-level line is beneath the top of atmosphere line throughout the wavelengths.

53. (3) eliminating or preventing access to houseplants until the child is over three years of age (Application) Children under three years of age often have access only to plants in the home; thus, eliminating house plants or preventing access to them would have the greatest impact. A healthy plant does not decrease its toxic effect. Pesticides increase toxicity. Eating commercial mushrooms is quite safe. Unless all parents of children read the published book and follow the suggestions, the book is ineffective.

54. (5) plant propagation text (Application) A plant propagation text would be primarily interested in indicating the best conditions for the plant's reproduction, not the survival or health of humans. All the other choices center on human survival.

55. (3) size (Analysis) The animals shown do not eat others that are larger than themselves. All other characteristics vary on the chain.

56. (3) there are more choices in what the animal can eat (Evaluation) The top animal has more choice in that it can eat all others on the chain. Being below another animal on the food chain does not increase the ease of being caught. Being easily seen is not an advantage. Not all animals at the top of a food chain can fly. An advantage of the bottom animal is not necessarily an advantage of the animal at the top.

57. **(4) The top must move faster than the middle or bottom.** (Analysis) In order for the top to reach the ground at the same time as the bottom, the top must move faster as the distance is farther. Since the distances are all different, the speeds needed to land at the same time must be different. Branches equaling trunk weight does not increase the speed for the top. The massive trunk has the shortest distance so it cannot move the fastest and land at the same time.

58. **(3) a change in the taste of water or the smell of the air** (Comprehension) Radon gas is odorless and tasteless. The human body cannot detect its presence. All other choices are presented in the passage as being directly related to radon gas.

59. **(4) the Environmental Protection Agency of the federal government** (Evaluation) The EPA of the federal government has no vested interest in making a profit. Its information is free and unbiased. The others may wish to make a profit or may give false or misleading information.

60. **(3) eating saturated fats with high cholesterol levels** (Analysis) Eating saturated fats with high cholesterol levels may be unhealthy but does not increase lung cancer risk. As noted in the passage, all other choices increase the risk in contaminated homes.

61. **(4) building a house over shale or granite rocks** (Application) To build a house over these rocks could endanger the health of the occupants. The other choices are wrong because they would decrease the risk of lung cancer. The question asks which would not decrease the risk.

62. **(5) a biocolloid** (Application) Blood is found in living organisms. No other answer choice is found in a live organism.

63. **(3) alcohol** (Comprehension) A tincture has alcohol as the dissolving medium. Water, mercury, gelatin, or sugar are not substances that define a tincture.

64. **(4) amalgam** (Analysis) Of the five mixtures described, only amalgam involves the use of metals.

65. **(1) an aerosol** (Application) Hair spray is a liquid in a can that is dispersed by a gas when the valve is opened. It does not involve metal, mercury, or alcohol, nor is it inside a living organism. Hair spray is quite different from hair gel, which is a liquid in a solid.

66. **(2) a gel** (Application) A gel is a liquid held in a solid. The hot water and fruit was held into a molded shape after the mixture cooled. There was no mention of gas, metal, mercury, living organisms, or alcohol.

SIMULATED GED SCIENCE TEST B

Pages 56–74

1. **(2) concerned** (Comprehension) If hypersensitive to bee stings, a person should remain calm but be concerned to seek medical attention. Giving up or being indifferent may result in not seeking medical attention. Harsh or excited behavior increases blood flow and distributes the poison more quickly.

2. **(1) rattlesnake bite to a dog** (Application) A rattlesnake bite puts a poisonous liquid in the dog, as does a bee sting. The fish bite, cat bite, heart attack, and accident are not caused by a poisonous liquid.

3. **(5) Molecular development is dependent on cellular formation.** (Comprehension) The chart is a series of levels by complexity. Cells are made of molecules, not molecules from cells. Cells are above organelles and are thus made up of organelles. Atoms at the innermost position make up all other levels. The biosphere is made of communities which are a lower level of complexity.

4. **(4) organisms, populations, communities, biosphere** (Analysis) Each item to the right must be more complex than and dependent upon all items to the left. Organisms, populations, communities, biosphere fits the diagram in least complex to the left and most complex to the right. All other options have one or more structures out of order.

5. **(2) the iron of an old bicycle combining with the oxygen in air to form rust** (Application) In rusting, the combining of oxygen and iron is slow and does not result in fire. All other options release noticeable heat or light quickly, resulting in fire.

6. **(3) maintenance of body temperature** (Analysis) Burning results in heat energy which is evidenced by body temperature. Heat or light, the evidence of burning fuels, is not indicated in any other option.

7. **(5) subway train** (Application) Most subway trains run by the electrification of a rail or overhead wires. All other options burn fuels and thus have engines.

8. **(4) establishing land grant colleges to increase agricultural productivity** (Evaluation) Increasing agricultural productivity increases the demand for power to operate the machines. All other options increase the supply.

9. **(4) flower pollination** (Application) To obtain seeds, flowers must be pollinated. Butterflies and moths assist flowers with cross-pollination. Butterfly reproduction would only be important if there were problems in obtaining a large enough

butterfly population to pollinate flowers. Chrysalis spinning and apple production would not generally be of interest to a flower seed grower. Butterflies and moths do not destroy worms.

10. **(5) silk production** (Comprehension) The production of silk by the silkworm moth is noted in the passage as the most important economic value of Lepidoptera. Spider webs are not spun by Lepidoptera. Elm tree pollination is accomplished by wind. Nectar is eaten by Lepidoptera, not turned to honey and stored. Some people collect moths and butterflies for their wings, but the economic market is small compared to the market for silk.

11. **(4) moth** (Analysis) Only moths construct cocoons. Butterflies form a hard chrysalis. Worms and Hymenoptera (bees and ants) do not form cocoons.

12. **(2) the possible destruction of oak trees** (Evaluation) The gypsy moth is responsible for much tree damage across the northeastern portion of the United States. New York is in the Northeast. Many people prefer to have butterflies in their gardens rather than bees which can sting. Nylon, made from petroleum, and silk do compete commercially. The silk-worm moth does not live in New York; it lives in China and India. If a convention of those who study and collect moths and butterflies was a failure, it is not likely that a town would take revenge on the moths and butterflies.

13. **(1) a dorsal fin** (Application) Dorsal is above and along the backbone; ventral is under the belly; lateral is to the side; caudal refers to the tail only; and posterior refers to a view from behind.

14. **(3) a posterior view** (Comprehension) Posterior refers to the rear; while anterior means the front. A lateral view is from the side; while a ventral view is the underneath or abdominal area. A cranial view is a view of the head.

15. **(4) the plant known as the Barbados lily is not botanically a lily** (Comprehension) The title indicates the items in the list are not lilies. All other items assume the plants are lilies.

16. **(2) identify the proper botanical classification for plants mistakenly called "lilies"** (Evaluation) The list gives the botanical names for plants that are mistakenly called lilies. Options 1, 3, and 4 assume the list is of lilies. Not all nonlilies are mistakenly called lilies; thus, the list does not classify all nonlilies.

17. **(1) Injuries are repaired and diseases are cured.** (Analysis) Taking aspirin does not cure a disease or repair an injury. Aspirin acts to block the body's natural actions to cure and repair. Aspirin relieves pain and lowers fever and blood flow. If healing or a cure takes place, it does so despite the aspirin, not because of the aspirin.

18. **(4) plastic cups** (Application) Plastic does not occur in nature and requires humans to synthesize it. Orange juice and cotton are plant products. Gold is a mineral, and leather is the skin of an animal.

19. **(3) they are able to synthesize complex substances** (Analysis) Modern societies are able to synthesize the complex substances that are required to return the body to a healthy state when the body malfunctions, is injured, or diseased. Having a high I.Q. does not mean I.Q. is necessarily used to obtain substances that prolong life. The ability to read and treat oneself does not make available the substances needed. Pain-killing drugs do not eliminate injury or the causes of stress. A baby will not remain healthy without needed substances.

20. **(5) a wide variety of foods to obtain essential vitamins and minerals by eating** (Application) Eating foods to obtain vitamins and minerals is a natural process. All other options produce substances not occurring naturally in plants, animals, or minerals.

21. **(4) time, size, mass, string thickness, and distance of pull** (Comprehension) The time, size, mass, string thickness, and distance of pull were the same for each pendulum. The length of the string was the variable being tested. In the scientific method, an experiment must have only one variable. All other factors must be controlled.

22. **(3) the length of the string** (Analysis) The length of the string was different for every pendulum. It is the only variable which indicates the factor being tested and the hypothesis that defines the test. Ball size, mass, and distance of pull were all controlled.

23. **(2) string length the same and vary the mass of the balls** (Analysis) Since only one factor can be tested at a time, the string lengths must be made the same and the mass of the balls varied. Varying distance of pull or ball size does not test mass. Varying both mass and size and leaving string length as in the illustration causes three variables.

24. **(3) dependent on the length of the string** (Analysis) The results show that the shorter the string length, the more swings per minute. Thus, the number of swings per unit time is dependent on string length. The mass of the balls and the thickness of string could not affect the swings as they were the same for each pendulum. The results were not constant, and since string length was varied, the results cannot be considered independent.

25. (5) roller coasters at amusement parks (Application) Roller coasters do not swing back and forth on a suspended rod as do the rubber tire swing, grandfather clock pendulum, wrecking balls, and the trapeze swing.

26. (4) the need for living things to breathe air (Application) Although we use the expression "air supports life," it does not hold the structure of the body to give it shape. The shape of breads, tires, balloons, and whipped cream is dependent on gas.

27. (2) length of cones (Comprehension) The key says Pinus coulteri has cones longer than 8 inches and that Pinus ponderosa has cones shorter than 5 inches.

28. (4) Pinus lambertiana (Application) If the tree has cones, it must be a conifer. If its cones hang down, it must be one of the Pinus trees. If its needles are in bunches of five, it must be a Pinus lambertiana.

29. (1) forester (Application) A forester manages forests of trees, and so would want to know the types of trees in a forest.

30. (4) another tribe building a village close by upstream (Evaluation) A new village upstream would result in pollution of the village's water supply. A new village downstream would not pollute areas upstream. The dye business is assigned to an appropriate area so as not to taint water for personal use. A government school or medical outpost would be forced to use the stream as the village dictates.

31. (3) contains less humus (Comprehension) The fertility of soil is dependent on the amount of humus. The fact that subsoil is further underground makes it more likely not to have as much humus, but position is not the factor that controls fertility. Particle size and ability to hold water are related to soil type, not fertility. Subsoil can be any mixture of soil types. The proportion of clay to sand controls drainage, not fertility.

32. (5) know both the soil characteristics and the soil needs of various crops (Analysis) Both the soil characteristics of a field and the soil needs of various plants must be considered to successfully produce crops. Many plants have particular soil needs; knowing only soil characteristics does not match the plant to the soil. Options 1, 2, and 3 are changes to be based on knowledge of soil characteristics and crop needs.

33. (1) increase the proportion of sand (Application) Water drains easily in sandy soil; thus, adding sand to the clay/loam soil would increase drainage. Increasing clay and decreasing sand would cause the soil to hold more water. Increasing the humus does not increase drainage, nor does the depth of the topsoil.

34. (4) Up means away from Earth. (Comprehension) Up means away from Earth. In the illustration, up refers to four different directions, so north, south, east, or west cannot define up. Down means toward the center of Earth and does not define up.

35. (4) objects are caught in the warp created by the mass of Earth (Comprehension) Earth's warp forms a hole in space that traps the objects in it. Air pressure of 14.7 lb./sq. in. is insufficient to keep objects moving at ordinary speeds from escaping into space if no warp existed. The passage and illustrations do not suggest that Earth's spin or magnetism is involved. People can live on any side of Earth.

36. (3) orbits the sun (Application) Earth orbits the sun because it is caught in the sun's warp and is not traveling fast enough to escape. The moon is caught in Earth's warp. The other options are not mentioned in the illustration or passage and are not effects of warps.

37. (4) an astronomer on Earth recording radio signals from radiating stars in the Andromeda Galaxy (Application) The recording of radio signals which are a form of energy would travel at 186,000 miles and not be affected by the warps of the sun, moon, or Earth. In all other options, an object will be moving through one or more warps, so fuel needed, determination of orbital paths, and the escape from and reentry into space of the object are all affected by the warps.

38. (2) a greater force than to escape Earth (Analysis) The greater the mass, the greater the warp. Therefore, the force to escape must be greater. A warp is a hole itself, and escaping the warp is not done by finding a hole in the hole. Escape is determined by speed, not the mass of the object attempting to escape.

39. (1) liquid (Comprehension) Liquids can flow and move more freely than solids such as rocks and soils. Both rocks and ocean water can be cold, vast, and deep. The color blue is not a reason for movement.

40. (4) 30 lb. (Analysis) One-sixth of 180 lb. equals 30 lb. The moon is smaller than Earth; therefore, the warp and gravity are also smaller, making the weight smaller, eliminating options 1, 2, and 3. Fifteen pounds would represent only 1/12 of the original weight.

41. (5) skin type and history of sunburns (Evaluation) The table indicates skin type and past sunburn experience as factors for choosing an SPF sunscreen product. A family history of cancer may not be related to skin cancer. Prevalence of cancer in an area may be due to factors other than sunburn. Color of skin, not eye color or hair color, is used to determine SPF number. The ease with which skin burns, not how long one is in the sun, determines an appropriate SPF number.

42. **(3) Exposure to the sun is now known to be the primary cause of skin cancer.** (Evaluation) Sunscreens help to prevent tanning and sunburn. Thus they are used to decrease the incidence of skin cancer. Knowledge of the active ingredient or the process of how skin tans or burns does not present a reason to choose to apply a sunscreen.

43. **(3) a characteristic that distinguishes a spider from an insect is the number of legs** (Analysis) Spiders and insects are listed on the graph and can be compared as having a difference in the number of legs. Birds have wings but not six legs. Only one sea creature is listed, so a generalization about all sea creatures cannot be made. The graph only lists the number of legs and does not give information on the complexity of entire bodies. The graph shows that ungulates (hoofed animals) have four legs but does not indicate that there are no other animals with four legs.

44. **(3) know that it is not an ungulate** (Application) Since all ungulates have four legs and the discovered animal has only two, it cannot be an ungulate. Animals other than birds, such as humans, can have two legs. The number of legs does not determine whether a creature can bite or live in water.

45. **(3) age and total number of offspring** (Application) All animals have an age, and the total number of offspring varies not only in number at delivery but at differing ages and at the cessation of reproduction. Hair, toes, ears, and tails are all frequently observed to distinguish between kinds of land animals.

46. **(3) tear production** (Comprehension) The production of tears is not directly related to the body's need for water but is an expression of emotion. Thirst, dry skin, cracked lips, and a dry mouth are all indicators of the body's need for more water.

47. **(5) wrap the feet in a thick, tight bandage** (Evaluation) Wrapping the feet would keep the area dark, warm, and moist and would not be effective in preventing favorable conditions for the survival of the fungus. Powder, open sandals, and absorbent socks all help decrease moisture. Open sandals also increase light. Spraying with an fungicide kills the fungus between the toes.

48. **(2) Sometimes the scientific solution for one problem presents another problem.** (Evaluation) The scientific solution to the farmer's problems presented a problem for the bay's fishing industry. No negative opinions against science, fertilizers, or the farmers were suggested. Using science did solve the original problem. The world's food problems were not discussed in the passage.

49. **(5) increased soil testing by the farmers** (Analysis) Soil testing increased as a result of state encouragement, not from use of chemicals. All other options resulted from the use of chemicals.

50. **(3) the farmers** (Comprehension) The chemicals benefited the farmers by increased income from increased production. All others suffered decreased income or increased prices.

51. **(4) discovery and production of electricity** (Evaluation) Extraction of metals by electrolysis was dependent upon the discovery and production of electricity. Although nuclear reactors and better furnace liners can aid in increasing the temperature, heating is not the primary method of extracting modern metals. The magnetism of iron did not contribute to the process of electrolysis. Space exploration came after the discovery of electrolysis.

52. **(3) rocks that contain a metal** (Comprehension) Ores are rocks that contain a metal. By definition, other options are incorrect and not indicated by the passage.

53. **(5) Mosses are found in all forests despite latitude and altitude.** (Analysis) Mosses are found at all levels, but the diagram does not support that conclusion, showing mosses only as surviving at high altitudes and latitudes. All others are correct and can be verified by the diagram.

54. **(1) both latitude and altitude are high** (Analysis) If either or both latitude and altitude are high, there are no forests. The diagram compares two causes of community type; both together or one alone can be the cause of a deciduous forest so that if latitude or altitude (either or both) are moderate, then a deciduous forest will result.

55. **(5) By mutation and variation of genetic material, many mosquitoes are no longer affected by DDT and other insecticides.** (Evaluation) Mosquitoes surviving a previous poison by mutation do not show that survival is dependent on the maintenance and balance of critical environmental factors as they survived an environmental imbalance. The examples of death, illness, or inability to reproduce when an imbalance occurs all support the statement about the importance of maintenance and balance of conditions for survival.

56. **(5) Some seeds are able to survive during long periods of dormancy.** (Analysis) Seeds from pyramid tombs germinating is evidence that some seeds can survive long periods of dormancy. Not all plants can be considered to perform as one example did. There is no reference in the statement to compare or evaluate animal reproduction.

57. **(5) arrangement of the atoms** (Analysis) The number of hydrogen and carbon atoms in isobutane and normal butane are the same; only the arrangement is different. Therefore, the difference in boiling points is likely due to the difference in arrangement.

58. **(5) blowing air through a bass tuba** (Application) Air vibrates when a tuba is played, but a tuba does not have a vibrating part. A vibrating violin string, drum skin, cymbal, and piano string are all a part of an instrument.

59. **(4) gears** (Comprehension) The rack, pinion, and end of the steering shaft form a gear mechanism. None of the other options are indicated by the diagram.

60. **(1) transfer steering movement to the pinion gear** (Analysis) The shaft connects the steering wheel to the pinion and enables the motion of the steering wheel to be transferred to the pinion. The steering arms do not move up and down. The housing holds the steering shaft in place rather than vice versa. The rods do not touch or join the steering shaft. The horn is not pictured in the diagram, so its relationship to the steering shaft is not known.

61. **(4) car mechanic** (Evaluation) The car mechanic must understand the operation of the mechanics of cars in order to diagnose breakdowns and initiate repair. Repair is not the major function of the individuals in all other options, so understanding is of lesser value to them.

62. **(4) swish or roll the liquid around the tongue prior to swallowing** (Analysis) Rolling the liquid around the tongue increases the amount and time of exposure to the receptor cells and intensifies the taste. Diluting decreases the concentration as would extra saliva. Swallowing quickly decreases the time on the receptor cells. Holding the nose decreases the inhaling of gases evaporating from the liquid, so the nose cannot heighten the pleasure of the taste.

63. **(5) the hard candy in a covered glass dish** (Application) The hard candy would evaporate very slowly, and the gases could not leave the dish because of the cover. Thus no chemicals from the candy could reach the nose. Gases evaporating from the cut grass, barbecue, burned beans, and banana move through the air and reach the nose.

64. **(2) aftershave lotion** (Application) Aftershave lotion contains chemicals that evaporate and move through the air. Their odors are mostly dependent on the chemicals in the lotion. Unless an individual is ill, fresh urine and perspiration have no smell. Only after microorganisms have processed the chemicals do the odors form. The odors of decay and rot also result from the processes of microorganisms.

65. **(5) A teenager exceeding the recommended daily water intake after playing football on a hot day** (Application) Since water is a common intake for all people, the teenager is not changing the composition or odor of the perspiration. Playing football may affect the amount but not the odor or composition. Meat eaters have a different odor than vegetarians due to the food eaten. Nurses learn to recognize the perspiration odors of certain diseases or infections. Some companies market separate men and women's deodorants as there are some sexual differences in pH and chemicals excreted. The smell of a used sheet reflects the perspiration of the individual sleeping on it.

66. **(3) place the spoon over the back portion of the tongue** (Application) Since the receptors for bitter taste are on the back of the tongue, placing the spoon on top of them prevents much of the liquid from reaching them. All other positions and sucking through a straw force the liquid over the bitter receptors.

Answer Sheet

GED Science Test

Name: ______________________ **Class:** ______________________ **Date:** ______

○ Simulated Test A ○ Simulated Test B

1 ①②③④⑤	**12** ①②③④⑤	**23** ①②③④⑤	**34** ①②③④⑤	**45** ①②③④⑤	**56** ①②③④⑤
2 ①②③④⑤	**13** ①②③④⑤	**24** ①②③④⑤	**35** ①②③④⑤	**46** ①②③④⑤	**57** ①②③④⑤
3 ①②③④⑤	**14** ①②③④⑤	**25** ①②③④⑤	**36** ①②③④⑤	**47** ①②③④⑤	**58** ①②③④⑤
4 ①②③④⑤	**15** ①②③④⑤	**26** ①②③④⑤	**37** ①②③④⑤	**48** ①②③④⑤	**59** ①②③④⑤
5 ①②③④⑤	**16** ①②③④⑤	**27** ①②③④⑤	**38** ①②③④⑤	**49** ①②③④⑤	**60** ①②③④⑤
6 ①②③④⑤	**17** ①②③④⑤	**28** ①②③④⑤	**39** ①②③④⑤	**50** ①②③④⑤	**61** ①②③④⑤
7 ①②③④⑤	**18** ①②③④⑤	**29** ①②③④⑤	**40** ①②③④⑤	**51** ①②③④⑤	**62** ①②③④⑤
8 ①②③④⑤	**19** ①②③④⑤	**30** ①②③④⑤	**41** ①②③④⑤	**52** ①②③④⑤	**63** ①②③④⑤
9 ①②③④⑤	**20** ①②③④⑤	**31** ①②③④⑤	**42** ①②③④⑤	**53** ①②③④⑤	**64** ①②③④⑤
10 ①②③④⑤	**21** ①②③④⑤	**32** ①②③④⑤	**43** ①②③④⑤	**54** ①②③④⑤	**65** ①②③④⑤
11 ①②③④⑤	**22** ①②③④⑤	**33** ①②③④⑤	**44** ①②③④⑤	**55** ①②③④⑤	**66** ①②③④⑤